INDEPENDENT COMPONENT ANALYSIS

THEORY AND APPLICATIONS

INDEPENDENT COMPONENT ANALYSIS

THEORY AND APPLICATIONS

by

TE-WON LEE
Computational Neurobiology Laboratory
The Salk Institute, La Jolla, California

SPRINGER-SCIENCE+BUSINESS MEDIA, B.V.

A C.I.P. Catalogue record for this book is available from the Library of Congress.

DOI 10.1007/978-1-4757-2851-4

Printed on acid-free paper

This book is dedicated to my parents Jeong-Bog Lee and Sun-Ja Kang, my sister Soon-Hie and my brother Yu-Won.

Contents

Abstract

Independent Component Analysis (ICA) is a signal processing method to extract independent sources given only observed data that are mixtures of the unknown sources. Recently, blind source separation by ICA has received attention because of its potential signal processing applications such as speech enhancement systems, telecommunications and medical signal processing.

This book presents theories and applications of ICA. Based on theories in probabilistic models, information theory and artificial neural networks several unsupervised learning algorithms are presented that can perform ICA. The seemingly different theories such as infomax, maximum likelihood estimation, negentropy maximization, nonlinear PCA, Bussgang algorithm and cumulant-based methods are reviewed and put in an information theoretic framework to unify several lines of ICA research (Lee et al., 1998a). An extension of the infomax algorithm of Bell and Sejnowski (1995) is presented that is able to blindly separate mixed signals with sub- and super-Gaussian source distributions (Girolami, 1997b; Lee et al., 1998b). The learning algorithms are furthermore extended to deal with the multichannel blind deconvolution problem. The use of filters allows the separation of voices recorded in a real environment (cocktail party problem). Several constraints in the ICA formulation such as the linear model assumption, the number of sensors and the low-noise assumption, are tackled with new methods. In particular, an overcomplete representation of the ICA formulation (Lewicki and Sejnowski, 1998c) whcih includes an additive noise model can be used to infer more sources than sensors.

The second part of the book presents signal processing applications of ICA to real-world problems. The ICA algorithm has been successfully applied to many biomedical signal processing problems such as the analysis of electroencephalographic (EEG) data (Makeig et al., 1997; Jung et al., 1998a) and functional magnetic resonance imaging (fMRI) data (McKeown et al., 1998b). Bell and Sejnowski (1997) suggested that independent components of natural scenes are edge filters. Those independent image components can be used as features in pattern classification problems such as visual lip-reading and face recognition systems. The ICA algorithm can be furthermore embedded in an expectation maximization framework with the goal to classify clusters of ICA models. This approach is an extension of the Gaussian mixture model for non-Gaussian priors. Results on classification benchmarks demonstrate that ICA cluster models can improve classification results.

ICA is a fairly new and a generally applicable method to several theoretical and practical challenges in signal processing. Successful results in EEG, fMRI, speech enhancement and face recognition systems demonstrate its power and its wide range of potential applications.

Preface

At a meeting held on April 13-16, 1986 in Snowbird Utah on Neural Networks for Computing Jeanny Herault and Christian Jutten (Herault and Jutten, 1986) contributed a research paper entitled "Space or time adaptive signal processing by neural network models". They presented a recurrent neural network model and a learning algorithm based on a version of the Hebb learning rule that, they claimed, was able to blindly separate mixtures of independent signals. They demonstrated the separation of two mixed signals and also mentioned the possibility of unmixing stereoscopic visual signals with four mixtures. This paper opens a remarkable chapter in the history of signal processing, a chapter that is hardly more than 10 years old.

The problem of source separation is an old one in electrical engineering and has been well studied; many algorithms exist depending on the nature of the mixed signals. The problem of blind source separation is more difficult since without knowledge of the signals that have been mixed, it is not possible to design appropriate preprocessing to optimally separate them. The only assumption made by Herault and Jutten was independence, but additional constraints are needed on the probability distribution of the sources. If one assumes, as is often done, that the source signals are Gaussian, then it is easy to show that this problem has no general solution. Subsequent research has shown that the best performance was obtained by the Herault-Jutten network when the source signals were sub-Gaussian (Cohen and Andreou, 1992); that is, for signals whose kurtosis was less than that of a Gaussian distribution.

In the neural network field, this network model was overshadowed at the time by the more popular Hopfield network, which would soon be eclipsed in popularity by the backpropagation algorithm for multilayer perceptrons. Nonetheless, a line of research was begun that only gradually made clear the true nature of the problem. As is often the case, what is important is not the specifics of the algorithm, but the way the problem is formulated. The general framework for independent component analysis introduced by Herault and Jutten is most clearly stated in Comon (1994). Within the signal processing community, a cornucopia of ever more sophisticated algorithms was developed based on cumulants, generalizing the third-order nonlinearity first used by Herault and Jutten.

By 1994 the forefront of the neural network field had moved from supervised learning algorithms to unsupervised learning. A fast and efficient ICA algorithm was needed that could scale up with the size of the problem at least as well as backpropagation, which by this time was being used on networks with over a million weights. Anthony Bell in my laboratory was working on an infomax (Linsker, 1992) approach to ICA. Tony's first results were obtained using Mathematica and a version of his algorithm that depended on inverting a matrix (Bell and Sejnowski, 1995). This was probably fortunate since the long pauses during convergence gave him ample time to think about the problem and to benefit from vigorous discussions with Nicol Schraudolph and Paul Viola, who at the time were sharing an office with a wonderful view of the Pacific Ocean. Both Nici and Paul were working on problems that involved estimating entropy gradients, so there was a keen competition to see whose algorithm would perform best. During this period, Tony collaborated by long-distance with Te-Won Lee, who at the time was visiting Carnegie- Mellon University, on blind source separation of acoustically recorded sound mixtures, taking into account time delays.

Amari (1997a) soon realized that the infomax ICA algorithm could be improved by using the natural gradient, which multiplies the gradient of the feedforward weight matrix $\mathbf{W}$ by a positive definite matrix $\mathbf{W}^T\mathbf{W}$, and speeds up the convergence by eliminating the matrix inversion. This improvement, which was independently discovered by Cardoso and Laheld (1996), allows infomax ICA to be scaled up and makes it a practical algorithm for a variety of real-world problems. However, the original infomax ICA algorithm was only suitable for super-Gaussian sources. Te-Won Lee realized that a key to generalizing the infomax algorithm to arbitrary non-Gaussian sources was to estimate moments of the source signals and to switch the algorithm appropriately. In collaboration with Mark Girolami, who had been working on similar algorithms in the context of projection pursuit, he soon developed an efficient extended version of the infomax ICA algorithm (Lee, Girolami and Sejnowski, 1998) that is suitable for general non-Gaussian signals. All of these developments are presented in this book with a clarity that makes them accessible to anyone with an interest in ICA.

Another important component of this book is the comparison between several different approaches that have been taken to blind source separation, which include maximum likelihood, Bussgang methods based on cumulants, and projections pursuit and negentropy methods. Te-Won Lee shows that these are all closely related to the infomax framework (Lee, Girolami, Bell and Sejnowski, 1998). Thus, a large number of researchers who have attacked ICA from a variety of different directions are converging on a common set of principles and, ultimately, a well understood class of algorithms. There is still much work that is left to do. It is still true as Herault and Jutten mention in their 1986 paper, "We cannot prove convergence of this algorithm because of nonlinearity of the adaptation law and nonstationarity of the signals." We still do not have an adequate explanation for why ICA does converge for so many problems, almost always to the same solutions, even when the signals were not derived from independent sources.

What makes this book especially valuable to the practitioner is that it also includes examples of several real-world applications. Although the blind separation of mixtures of prerecorded signals is a useful benchmark, a more challenging problem is to apply ICA to recordings of real-world signals for which the underlying sources, if any, are unknown. An important example is the application of extended infomax ICA to electroencephalographic (EEG) recordings of scalp potentials in humans. The electrical signals originating from the brain are quite weak at the scalp, in the microvolt range, and there are larger artifactual components arising from eye movements and muscles. It has been a difficult challenge to eliminate these artifacts without altering the brain signals. ICA is ideally suited to this task, since the brain and the scalp are good volume conductors and to a good approximation, the recordings are different linear mixtures of the brain signals and the artifacts. The extended infomax ICA algorithm has proven to be the best method yet for separating out these artifacts, which include sub-Gaussian sources such as 60 Hz line noise and blinks, from the brain signals, which are generally super-Gaussian (Jung et al., 1998a). The future of this algorithm looks quite bright for biomedical applications, including the analysis of extremely large datsets from functional Magnetic Resonance Imaging (fMRI) experiments (McKeown et al., 1998b).

ICA can be applied to many problems where mixtures are not orthogonal and the source signals are not Gaussian. Most information bearing signals have these characteristics. There are many interesting theoretical problems in ICA that have yet to be solved and there are many new applications, such as data mining, that have yet to be explored. The appearance of this book marks an important milestone in the maturation of ICA research. The theoretical framework developed here should provide a strong foundation for future research and applications.

TERRENCE SEJNOWSKI

Acknowledgments

This book was made possible with the enormous support of many people. Most particularly Tony Bell and Terry Sejnowski gave guidance and advice from the beginning of my studies in ICA until now. The scientific collaboration and discussions I had with them are invaluable. Tony inspired me to think about ideas so thoroughly that he changed an engineering reasoning into a scientific mind. Working with Terry was like walking with an encyclopedia. His broad knowledge was fascinating and although he was constantly collaborating with many other researchers at the same time he was always accessible until late after midnight.

I was privileged to stay at the Salk Institute, the computational neurobiology lab. (CNL), and to work with the leading researchers in biomedical signal processing: Tzyy-Ping Jung, Martin McKeown and Scott Makeig. The combined force of deep theoreticians and responsible practical researchers formed the dynamite *symmetrically doubled*-ICA team.

Special thanks go out to some international experts in ICA: Mark Girolami, Michael Lewicki, Jean-François Cardoso, Russ Lambert and Bert-Uwe Köhler for collaborations, superb discussions and comments. In general, this book would have been impossible without the help of so many researchers working in this field. I appreciate their valuable discussions and comments during conferences and via email and fax.

I am indebted to many organizations for their financial support and general guidance: Max-Planck-Gesellschaft, AG fehlertolerantes Rechnen, Carnegie Mellon University, Alex Waibel's speech group, Daimler-Benz fellowship, Deutscher Akademischer Austauschdienst (DAAD) fellowship and Deutsche Forschungsgemeinschaft (DFG) postdoc fellowship.

Individual thanks go out to alumni of Club Thesis (Olivier, Michael, Marni, and K.T.) for pushing the work, and many other CNL-ers for making my stay at the Salk as pleasant as possible. Special thanks go out to Tim for constant support from Berlin and my godly woman Jee Hea for her caring love and *wise* thoughts.

List of Figures

List of Tables

Abbreviations and Symbols

Abbreviations

ARMA	autoregressive moving average
BCM	Bienenstock, Copper and Moore theory of synaptic plasticity
BSS	blind source separation
CDMA	code division multiple access
c.d.f.	cumulative density function
c.g.f.	cumulant generating function
DCA	dynamic component analysis
EEG	electroencephalographic
EOG	electrooculographic
ERP	event-related potentials
FIR	finite impuls response
fMRI	functional magnetic resonance imaging
ICA	independent component analysis
IIR	infinite impuls response
iff	if and only if
infomax	principle of information maximization preservation
KL	Kullback-Leibler
MAP	maximum a posteriori
MEG	magnetoencephalography
m.g.f.	moment generation function
MLE	maximum likelihood estimation
PCA	principal component analyis
PET	positron emission tomography
p.d.f.	probability density function
QPSK	quadrature phase shift keying
RV	random variable
SOM	self-organizing maps
SOFM	self-organizing feature maps
SVD	singular value decomposition

TDD	time-delayed decorrelation

Symbols

Vectors, matrices and high-dimensional matrices are written in boldface. Usually vectors are lower-case type and matrices are higher-case type.

$\mathbf{A}$	mixing matrix with elements a_{ij}
$\mathbf{A}(z)$	mixing matrix of filters with elements $A_{ij}(z)$
A_x	set of possible outcomes $\{a_1, a_2, \cdots\}$
$\mathbf{a}$	basis functions of $\mathbf{A}$ (columns of $\mathbf{A}$)
$\mathbf{b}$	bias term
$\mathbf{C}$	covariance matrix
$C(.)$	cost function
c_i	i-th order cumulant
$\mathbf{D}$	diagonal matrix
$D(.\|\|.)$	Kullback-Leibler divergence
D_{ij}	mixing time-delay from channel i to channel j
d_{ij}	unmixing time-delay from channel i to channel j
d_i, r_i	inverse of scale d_i and slope $r_i$parameters for inverting the nonlinear transfer function
$\mathbf{E}$	matrix of which columns are eigenvectors of the covariance matrix
$E\{.\}$	expected value
$e(.)^2$	mean-squared error
$g(.)$	nonlinear activation function
$f(.)$	an arbitrary function
f_{jk}	coefficients of the polynomial expansion of the transfer function nonlinearity
$H(.)$	differential entropy
h_i	orthogonal hermite polynomials
$h(.)$	nonlinear inverse transfer function
h_{ik}	coefficients of the polynomial expansion of the flexible nonlinearity
$\mathbf{i_s}$	information symbols
$\mathbf{I}$	identity matrix
$I(.)$	mutual information
$J(.)$	negentropy
$\mathbf{J}$	Jacobian matrix
K	number of classes
k_i	switching moments for sub- and super-Gaussian densities
$\mathbf{K}$	diagonal matrix with elements k_i
$L(.)$	log-likelihood function
M	number of sources
m_i	i-th order central moment
N	number of sensors
N_X	number of elements in X

$N(a,b)$	Gaussian distribution with mean a and variance b
n	white Gaussian noise signal
$O(.)$	objective function
$\mathbf{P}$	performance matrix
$P(.)$	Probability
$p(.)$	probability density function
P	P^{th}-order of the nonlinearity mixing model
Q	Q^{th}-order of the flexible nonlinearity unmixing model
$\mathbf{R}$	permutation matrix
R_n	Lagrange's form of the remainder of a Taylor series expansion
$\mathbf{s}$	independent source signals $\mathbf{s} = [s_1(t), s_2(t), \cdots, s_M(t)]^T$
$\mathbf{t}$	linear mixed signals in the nonlinear mixing model
t	time index
$T(.)$	target function
$\mathbf{u}$	unmixed signals $\mathbf{u} = [u_1(t), u_2(t), \cdots, u_N(t)]^T$
$\hat{\mathbf{u}}$	estimated unmixed signals
$\mathbf{V}$	whitening matrix
$\mathbf{w}$	weight vector
$\mathbf{W}$	weight matrix
$\hat{\mathbf{W}}$	weight matrix for the feedback system
X	random variable with a limited number of possible outcomes
x	continuous random variable
$\mathbf{x}$	observed vector of continuous signals $\mathbf{x} = [x_1(t), x_2(t), \cdots, x_N(t)]^T$
$\mathbf{X}$	ICA mixture model:
	K classes of observed vectors $\mathbf{X} = [\mathbf{x}_1, \cdots, \mathbf{x}_k, \cdots, \mathbf{x}_K]^T$
Y	random variable with a limited number of possible outcomes
$\mathbf{y}$	neural network outputs $\mathbf{y} = [y_1(t), y_2(t), \cdots, y_N(t)]^T$
$\mathbf{z}$	signals after the nonlinear unmixing $\mathbf{z} = [z_1(t), z_2(t), \cdots, z_N(t)]^T$
FFT(.)	fast fourier transform
IFFT(.)	inverse fast fourier transform
tr(.)	trace function
$\Delta(z)$	determinant of the matrix of filters in the frequency domain
δ_i,σ_i	scale δ_i and slope σ_i parameters of the nonlinear transfer function
ϵ	learning rate
$\mu_x(i)$	i-th order moment of x
σ^2	variance
ω	Model
Ω	set of models
$\Phi(.)$	cumulant generating function
$\psi(.)$	moment generating function
$\Psi(.)$	higher-order polynomial terms of the Edgeworth expansion
$\varphi(.)$	score function
ρ_{ij}	correlation coefficient between signals x_i and x_j.

INTRODUCTION

"Begin at the beginning," the King said, gravely, " and go on till you come to the end; then stop."
Lewis Carroll

A new star is born: ICA

Recently, blind source separation by Independent Component Analysis (ICA) has received attention because of its potential applications in signal processing such as in speech recognition systems, telecommunications and medical signal processing.

The goal of blind source separation (BSS) is to recover independent sources given only sensor observations that are linear mixtures of independent source signals. The term *blind* indicates that both the source signals and the way the signals were mixed are unknown. Independent Component Analysis (ICA) is a method for solving the blind source separation problem. It finds a linear coordinate system (the unmixing system) such that the resulting signals are statistically independent. In contrast to correlation-based transformations such as Principal Component Analysis (PCA), ICA not only decorrelates the signals (2nd-order statistics) but also reduces higher-order statistical dependencies. In other words, *ICA is a method for finding a linear non-orthogonal co-ordinate system in any multivariate data. The directions of the axes of this co-ordinate system are determined by both the second and higher order statistics of the original data. The goal is to perform a linear transform which makes the resulting variables as statistically independent from each other as possible.*

Two different research communities have considered the analysis of independent components. On one hand, the study of separating mixed sources observed in an array of sensors has been a classical and difficult signal processing problem. The seminal work on blind source separation was by Herault and Jutten (1986) where they introduced an adaptive algorithm in a simple feedback architecture that was able to separate several unknown independent sources. Their approach has been further developed by Jutten and Herault (1991), Karhunen and Joutsensalo (1994), and Cichocki et al. (1994). Comon (1994) elaborated the concept of independent component analysis and proposed cost functions related to the approximate minimization of mutual information between the sensors.

In parallel to blind source separation studies, unsupervised learning rules based on information theory were proposed by Linsker (1992). The goal was to maximize the mutual information between the inputs and outputs of a neural network. This approach is related to the principle of redundancy reduction suggested by Barlow (1961) as a coding strategy in neurons. Each neuron should encode features that are as statistically independent as possible from other neurons over a natural ensemble of inputs; decorrelation as a strategy for visual processing was explored by Atick (1992). Nadal and Parga (1994) showed that in the low-noise case, the maximum of the mutual information between the input and output of a neural network implied that the output distribution was factorial; that is, the multivariate probability density function (p.d.f.) can be factorized as a product of marginal p.d.f.s. Roth and Baram (1996) and Bell and Sejnowski (1995) independently derived stochastic gradient learning rules for this maximization and applied them, respectively, to forecasting, time series analysis, and the blind separation of sources. Bell and Sejnowski (1995) put the blind source separation problem into an information-theoretic framework and demonstrated the separation and deconvolution of mixed sources. Their adaptive methods are more plausible from a neural processing perspective than the cumulant-based cost functions proposed Comon (1994). A similar adaptive method for source separation was proposed by Cardoso and Laheld (1996).

Other algorithms for performing ICA have been proposed from different viewpoints. Maximum Likelihood Estimation (MLE) approaches to ICA were first proposed by Gaeta and Lacoume (1990) and elaborated by Pham et al. (1992). Pearlmutter and Parra (1996), MacKay (1996) and Cardoso (1997) showed that the infomax approach of Bell and Sejnowski (1995) and the maximum likelihood estimation approach are equivalent. Girolami and Fyfe (1997c) motivated by information-theoretic indices for Exploratory Projection Pursuit (EPP) used marginal negentropy as a projection index and showed that kurtosis-seeking projection pursuit will extract one of the underlying sources from a linear mixture. A multiple output EPP network was developed to allow full separation of all the underlying sources (Girolami and Fyfe, 1997b). Nonlinear PCA algorithms for ICA which have been developed by Karhunen and Joutsensalo (1994), Xu (1993) and Oja (1997) can also be viewed from the infomax principle since they approximately minimize the sum of squares of the fourth-order marginal cumulants (Comon, 1994) and therefore approximately minimize the mutual information of the network outputs (Girolami and Fyfe, 1997d). Bell and Sejnowski (1995) have pointed out a similarity between their infomax algorithm and the Bussgang algorithm in signal processing and Lambert (1996) elucidated the connection between three different Bussgang cost functions. Lee et al. (1998a) show how the Bussgang property relates to the infomax principle and how all of these seemingly different approaches can be put into a unifying framework for the source separation problem based on an information theoretic approach.

The original infomax learning rule for blind separation by Bell and Sejnowski (1995) was suitable for super-Gaussian sources. An extension of the infomax algorithm of Bell and Sejnowski (1995) is presented in Lee et al. (1998b) that is able to blindly separate mixed signals with sub- and super-Gaussian source distributions. This was achieved by using a simple type of learning rule first derived by Girolami (1997b) by choosing negentropy as a projection pursuit index. Parameterized probability distributions that have sub- and super-Gaussian regimes were used to derive a general learning rule that preserves the simple architecture proposed by Bell and Sejnowski (1995), is optimized using the natural gradient by Amari (1998), and uses the stability analysis of Cardoso and Laheld (1996) to switch between sub- and super-Gaussian regimes.

Extensive simulations have been performed to demonstrate the power of the learning algorithm. However, instantaneous mixing and unmixing simulations are *toy* problems and the challenge lies in dealing with real world data. Makeig et al. (1996) have applied the original infomax algorithm to EEG and ERP data showing that the algorithm can extract EEG activations and isolate artifacts. Jung et al. (1998a) show that the extended infomax algorithm is able to linearly decompose EEG artifacts such as line noise, eye blinks, and cardiac noise into independent components with sub- and super-Gaussian distributions. McKeown et al. (1998b) have used the extended ICA algorithm to investigate task-related human brain activity in fMRI data. By determining the brain regions that contained significant amounts of specific temporally independent components, they were able to specify the spatial distribution of transiently task-related brain activations. Other potential applications may result from exploring independent features in natural images. Bell and Sejnowski (1997) suggest that independent components of natural scenes are edge filters. The filters are localized, mostly oriented and similar to Gabor like filters. The outputs of the ICA filters are sparsely distributed. Bartlett and Sejnowski (1997) and Gray et al. (1998) demonstrate the successful use of the ICA filters as features in face recognition tasks and lipreading tasks respectively. In a similar manner, Bell and Sejnowski (1996) applied the infomax algorithm to learning higher-order structure of a natural sound.

For these applications, the instantaneous mixing model may be appropriate because the propagation delays are negligible. However, in real environments substantial time-delays may occur and an architecture and algorithm is needed to account for the mixing of time-delayed sources and convolved sources. The multichannel blind source separation problem has been addressed by Yellin and Weinstein (1994) and Nguyen-Thi and Jutten (1995) and others based on 4th-order cumulants criteria. An extension to time-delays and convolved sources from the infomax viewpoint using a feedback architecture was developed by Torkkola (1996a). Lee et al. (1997a) extended the blind source separation problem to a full feedback system and a full feedforward system. The feedforward architecture allows the inversion of non-minimum phase systems. In addition, the rules are extended using polynomial filter matrix algebra in the frequency domain (Lambert, 1996). The proposed method can successfully separate voices and music recorded in a real environment. Lee et al. (1997b) showed that the recognition rate of an automatic speech recognition system was increased after separating the speech signals.

Since ICA is restricted and relies on several assumptions researchers have started to tackle a few limitations of ICA. One obvious but non-trivial extension is the nonlinear mixing model. In (Hermann and Yang, 1996; Lin and Cowan, 1997; Pajunen, 1996) nonlinear components are extracted using self-organizing-feature-maps (SOFM). Other researchers (Burel, 1992; Lee et al., 1997c; Taleb and Jutten, 1997; Yang et al., 1997; Hochreiter and Schmidhuber, 1998) have used a more direct extension to the previously presented ICA models. They include certain flexible nonlinearities in the mixing model and the goal is to invert the linear mixing matrix as well as the nonlinear mixing. More recently, Hochreiter and Schmidhuber (1998) have proposed low complexity coding and decoding approaches for nonlinear ICA. Another limitation is the under-determined problem in ICA, i.e. having less sensors than sources. Lee et al. (1998c) demonstrated that an overcomplete representation (Lewicki and Sejnowski, 1998b) of the data can be used to learn non-square mixing matrices and to infer more sources than sensors. The overcomplete framework also allows additive noise in the ICA model and can therefore be used to separate noisy mixtures.

There is now a substantial amount of literature on ICA and BSS. Reviews of the different theories can be found in Cardoso and Comon (1996); Cardoso (1998b); Lee et al. (1998a) and Nadal and Parga (1997). Several neural network learning rules are reviewed and discussed by Karhunen (1996); Cichocki and Unbehauen (1996) and Karhunen et al. (1997a).

ICA is a fairly new and a generally applicable method to several challenges in signal processing. It reveals a diversity of theoretical questions and opens a variety of potential applications. Successful results in EEG, fMRI, speech recognition and face recognition systems indicate the power and optimistic expectations in the new paradigm.

Organization of the book

This book is partitioned into Theory and Applications of ICA. The theory part of ICA includes(basic theory (chapter 1), ICA (chapter 2), unifying approach for ICA (chapter 3), multichannel deconvolution (chapter 4), overcomplete ICA (chapter 5) and nonlinear ICA (chapter 6).

- Chapter 1 starts with an introduction to Bayesian probability theory, information theory, artificial neural networks and higher-order statistics. Only some basics and some properties that are needed are recalled to further derive and explain the methods and algorithms performing ICA.

- Chapter 2 states the ICA problem and explains why decorrelation-based methods fail to separate sources and why higher-order methods are needed to solve this problem. The unsupervised learning algorithm by Bell and Sejnowski (1995) is analyzed that is able to blindly separate mixed sources with super-Gaussian distributions. An extension of this algorithm is presented that is able to blindly separate mixed signals with sub- and super-Gaussian source distributions (Girolami, 1997b; Lee et al., 1998b).

- Chapter 3 presents different theories recently proposed for ICA and show how they lead to the same iterative learning algorithm for blind separation of mixed independent sources. Those seemingly different theories are reviewed and put in an information theoretic framework to unify several lines of research.

- Chapter 4 deals with time-delayed and convolved sources, the multichannel blind deconvolution problem. Learning algorithms for the feedforward and feedback architecture are presented. These methods is applied to multisensory recordings and is able to separate voices and music recorded in real environments.

- Chapter 5 presents ICA results using an overcomplete representation (Lewicki and Sejnowski, 1998b). This generalization includes an additive noise model and allows the separation of more sources than mixtures.

- Chapter 6 tries to eliminate several constraints in the standard ICA formulation. The linear model is generalized by simple nonlinear mixing models for nonlinear ICA. A set of learning algorithms are derived and verified in simulations.

The applications part of ICA includes ICA for biomedical signal processing (chapter 7), feature extraction (chapter 8) and ICA for unsupervised classification (chapter 9).

- Chapter 7 presents examples on how ICA can be used to (1) isolate artifacts in EEG recordings and (2) to detect transiently activated brain signals in fMRI experiments.

- Chapter 8 demonstrates how ICA can extract features from natural images. Those features can be used in recognition systems such as face recognition and lipreading systems to improve overall performance.

- Chapter 9 shows how the ICA algorithm can be further embedded in an expectation maximization framework with the goal to classify mixtures of ICA models. This approach is an extension of the Gaussian mixture model for non-Gaussian priors. Results on several classification benchmarks demonstrate that ICA cluster models can improve many classification results.

Chapter 10 gives conclusions by summarizing the main results and discussing future challenges in ICA research.

I Independent Component Analysis: Theory

1 BASICS

Everything should be made as simple as possible,
but not simpler.
Albert Einstein

1.1 OVERVIEW

This chapter is an introduction to the basics of Bayesian probability theory, information theory, artificial neural networks and statistical signal processing. The material presented here was selected from several textbooks. The goal of this chapter is to cover the basic notions and terminologies used throughout the thesis. It is not intended to give a broad understanding of the theories but rather to recall some definitions and their relations to each other.

Bayesian probability theory is summarized by giving definitions for maximum a posteriori estimation and maximum likelihood estimation. Information theory is divided into basic definitions, the notion of differential entropy and the maximum entropy property. The section on artificial neural networks focuses on unsupervised learning algorithms for data analysis. For example, a simple single neuron can find directions of maximal variance in a data set using Oja's learning rule. Its extension to the generalized Hebbian algorithm can perform the well known principal component analysis (PCA). Another example for unsupervised learning algorithms is the information maximization preservation principle (infomax principle) proposed by

Linsker (1989) in a single layer feedforward networks and its relation to redundancy reduction. The last section deals with a brief summary about higher-order statistics. Important properties and definitions of higher-order moments and higher-order cumulants are presented.

There are several books that cover the summarized theories in much detail and precision. For example, David MacKay's book (MacKay, 1998) on **Information Theory, Inference and Learning Algorithm** is a good textbook that covers the summarized theories in great detail [1]. A good book on information theory is *Elements of Information Theory* by Cover and Thomas (1991). On neural networks, books by Haykin (1994b), Bishop (1995), Rojas (1996) and Hertz et al. (1991) give excellent overviews and explanations of neural networks. Bayesian probability theory is well explained by Box and Tiao (1992) and Bayesian theory from a neural perspective is presented by MacKay (1995). General statistics are in books by Stuart and Ord (1987) and Papoulis (1990). Cichocki and Unbehauen (1994) and Kosko (1992) present statistical signal processing applications using neural networks.

This chapter is organized as follows: In section 1.2 we present basic definitions for Bayesian probability theory, section 1.3 gives an introduction to information theory, section 1.4 presents some unsupervised learning algorithms in artificial neural networks and section 1.5 gives basic definitions and properties of higher-order moments and cumulants. The chapter closes with a brief discussion about some common relationships between the summarized theories.

1.2 BAYESIAN PROBABILITY THEORY

Bayesian reasoning provides a probabilistic approach to estimation and inference. It is based on the assumption that the quantities of interest are governed by probability distributions, and that optimal decisions can be made by reasoning about these probabilities together with the data. Bayesian approaches provide a framework for learning algorithms that manipulate probabilities directly as well as for learning algorithms that do not explicitly manipulate probabilities. Features of Bayesian approaches include the following facts:

- Bayesian methods can provide a standard of optimal decision making to compare different practical methods.
- Prior knowledge can be combined with the observed data to determine the final probability of a model.
- Each observed training example incrementally decreases or increases the estimated probability that a hypothesis or model is correct. This provides a more general method than methods that eliminate a model if it is inconsistent with any single observation.

In learning, a common interest is the determination of the best model from some model space given the observed data. In other words, what is the most probable model ω among a set of M models Ω given a data sample X plus any prior knowledge about the various models. Here, Bayes theorem provides a way to calculate the probability of a model based on its prior probability and the observed data itself. It is defined for any model ω as

$$P(\omega_i|X) = \frac{P(X|\omega_i)P(\omega_i)}{P(X)} \tag{1.1}$$

where $P(X|\omega)$ is the conditional probability or posterior probability of the observation X given the model ω, $P(\omega)$ is the probability of the model (prior probability of ω) and similarly $P(X)$ is the prior probability that the data X is observed. $P(\omega|X)$ is called the posterior probability of ω because it reflects the confidence that ω holds after the data X has been observed. In contrast to the prior $P(\omega)$, the posterior probability $P(\omega|X)$ reflects the influence of X whereas $P(\omega)$ is independent of X.

In many learning situations, there is a set of models $\Omega = \{\omega_1, \cdots, \omega_i, \cdots, \omega_M\}$ and the task is to find the most probable model $\omega_i \in \Omega$ given the observed data X. The maximally probable model is called a Maximum A Posteriori (MAP) hypothesis. The MAP hypothesis can be determined using the Bayes theorem.

$$\begin{aligned} \omega_{MAP} &= \arg\max_{\omega\in\Omega} P(\omega|X) \\ &= \arg\max_{\omega\in\Omega} \frac{P(X|\omega)P(\omega)}{P(X)} \\ &= \arg\max_{\omega\in\Omega} P(X|\omega)P(\omega). \end{aligned} \tag{1.2}$$

In the final step of eq.1.2 $P(X)$ is dropped because it is a constant independent of ω.

In some cases, it can be assumed that every model is equally probable a priori ($P(\omega_i) = P(\omega_j)$ for all i, j). Then, eq.1.2 further simplifies and it is sufficient to consider only the term $P(X|\omega)$ to find the most probable model. $P(X|\omega)$ is also called a Maximum Likelihood (ML) hypothesis

$$\omega_{ML} = \arg\max_{\omega\in\Omega} P(X|\omega) \tag{1.3}$$

where ω_{ML} is the most probable model among other models $\omega_i \in \Omega$ describing the observation X.

1.3 INFORMATION THEORY

Information is closely related to randomness or surprisal of an outcome. For example, in a coin tossing experiment that has normally two equiprobable outcomes: head and tail, the information that can be gained from each experiment is that the probability of observing a head is 0.5. This process is still random and it may be surprising that a head is observed although a tail was guessed. However, if it is known that someone made a fake coin with both sides of the coin showing heads the outcome of each experiment would not be surprising because there is nothing random. In fact, this deterministic experiment provides *no* information and its outcome is always certain. A way to measure the randomness of the coin tossing experiment depending on the

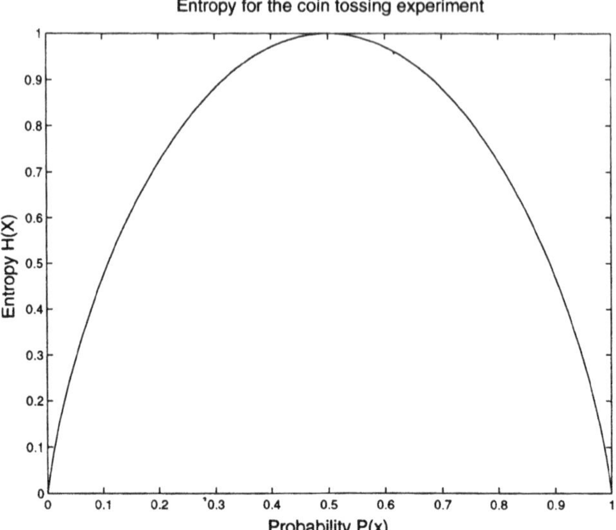

Figure 1.1. Entropy as a function of the probability P(x) for the coin tossing experiment. Maximum entropy is achieved for equiprobable outcomes and the entropy is zero for deterministic outcomes P(x)=1 and P(x)=0.

bias of the coin reveals whether the result is equiprobable, slightly biased towards one side or even fake. Shannon (1948) proposed entropy $H(X)$ as an appropriate measure. Entropy is defined as

$$H(X) = \sum_{x \in A_X} P(x) \log \frac{1}{P(x)}, \tag{1.4}$$

where the ensemble X is a random variable x with a set of possible outcomes, $A_x = \{a_1, a_2, ...\}$ and $\sum_{a_i \in A_x} P(x = a_i) = 1$. For $P(x) = 0$ the entropy is zero by definition. $H(X)$ is always greater or equal zero. Note that throughout this section and as opposed to the other sections x is not a continuous variable but a limited set of possible outcomes of a random variable X. In the coin tossing experiment $i = 2$ and $H(X)$ can be plotted as the function of the probability $P(x)$.

Notice that in figure 1.1 the curve is symmetrical, and rises to a maximum when the two symbols (head, tail) are equally likely. It falls towards zero once the other symbol becomes dominant and $H(X)$ is zero when the probability of the symbols are deterministic. In general, for a given number of symbols (a_i's) the entropy has its largest value only when the symbols are equally probable. The resulting real value (code length) for $H(X)$ can be described in bits when the log of base 2 is used and nats when the natural log is used.

The joint entropy of two random variables X and Y is defined as

$$H(X,Y) = \sum_{x,y \in A_X A_Y} P(x,y) \log \frac{1}{P(x,y)}. \tag{1.5}$$

It is additive for independent variables

$$H(X,Y) = H(X) + H(Y) \text{ iff } P(x,y) = P(x)P(y). \tag{1.6}$$

Eq.1.6 equals eq.1.5 as follows

$$\begin{aligned}
H(X,Y) &= \sum_{x,y \in A_X A_Y} P(x)P(y) \log \frac{1}{P(x)P(y)} \\
&= \sum_{x,y \in A_X A_Y} P(x)P(y) \log \frac{1}{P(x)} + \sum_{x,y \in A_X A_Y} P(x)P(y) \log \frac{1}{P(y)} \\
&= \sum_{x \in A_X} P(x)(\sum_{y \in A_Y} P(y)) \log \frac{1}{P(x)} + \sum_{y \in A_Y} P(y)(\sum_{x \in A_X} P(x)) \log \frac{1}{P(y)} \\
&= \sum_{x \in A_X} P(x) \log \frac{1}{P(x)} + \sum_{y \in A_Y} P(y) \log \frac{1}{P(y)} \\
&= H(X) + H(Y).
\end{aligned} \tag{1.7}$$

The conditional entropy of X given $y = b_k$ is the entropy of the conditional probability distribution $P(x|y = b_k)$

$$H(X|y = b_k) \equiv \sum_{x \in A_X} P(x|y = b_k) \log \frac{1}{P(x|y = b_k)}. \tag{1.8}$$

The conditional entropy of X given Y is the average over y of the conditional entropy of X given y and therefore measures the average uncertainty that remains about x when y is known

$$H(X|Y) = \sum_{x,y \in A_X A_Y} P(x,y) \log \frac{1}{P(x|y = b_k)}. \tag{1.9}$$

The mutual information between X and Y is the sum of marginal entropies minus the joint entropy.

$$\begin{aligned}
I(X;Y) &\equiv H(X) + H(Y) - H(X,Y) \\
&= H(X) - H(X|Y).
\end{aligned} \tag{1.10}$$

It is always greater than zero and it measures the reduction in uncertainty about x that results from learning the value of y.

The joint entropy $H(X,Y)$, the conditional entropy $H(X|Y)$ and the marginal entropy $H(X)$ or $H(Y)$ are related as follows

$$H(X,Y) = H(X) + H(Y|X) = H(Y) + H(X|Y). \tag{1.11}$$

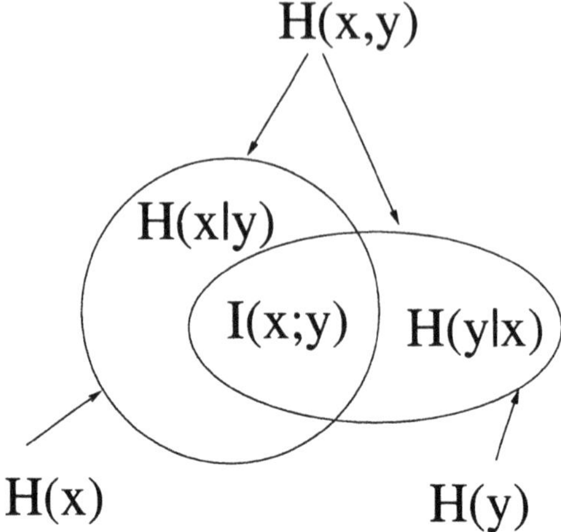

Figure 1.2. Marginal entropies $H(X)$ and $H(Y)$, joint entropy $H(X,Y)$, conditional entropy $H(X|Y)$ and the mutual information $I(X;Y)$.

The relative entropy also known as Kullback-Leibler divergence (KL divergence) between two probability distributions $P(x)$ and $Q(x)$ that are defined over the same alphabet A_X is

$$D(P\|Q) \equiv \sum_x P(x) \log \frac{P(x)}{Q(x)}. \tag{1.12}$$

The mutual information $I(X;Y)$ is a special form of the relative entropy or KL divergence and measures the distance between the joint entropy and the product distribution

$$I(X;Y) = D(P(x,y)\|P(x)P(y)). \tag{1.13}$$

1.3.1 Differential Entropy

So far the entropy terms were based on discrete probability distributions where the number of samples is finite, i.e. a random variable has a finite number of symbols. The entropy terms for a continuous random variable x is called differential entropy. It is defined as

$$h(x) \equiv \int_{-\infty}^{\infty} p(x) \log \frac{1}{p(x)} dx, \tag{1.14}$$

where $p(x)$ is the probability density function (p.d.f.) of a continuous random variable x. Differential entropy has the same properties and definitions as entropy in terms of its relation to conditional entropy, joint entropy and mutual information. Therefore, the above equations (eq.1.6, eq.1.8 and eq.1.10) are equivalent for differential entropies. One distinct difference however, is that differential entropies cannot be described in code length because differential entropies can take on negative values. In fact, the differential entropy can be proven to be $-\infty$ (Cover and Thomas, 1991; Haykin, 1994b)

$$H(X) = -\lim_{\delta x \to 0} \sum_{i=-\infty}^{\infty} p(x_i)\delta x \log(p(x_i)\delta x) \tag{1.15}$$

$$= -\lim_{\delta x \to 0} \left[\sum_{i=-\infty}^{\infty} p(x_i) \log(p(x_i))\delta x + \log \delta x \sum_{i=-\infty}^{\infty} p(x_i)\delta x \right]$$

$$= -\int_{-\infty}^{\infty} p(x) \log p(x) dx - \lim_{\delta x \to 0} \log \delta x \int_{-\infty}^{\infty} p(x) dx. \tag{1.16}$$

Eq.1.15 is the definition of entropy when the quantization of x converges to zero. The first term of the right hand side of eq.1.16 is by definition the differential entropy. Rewriting eq.1.16 in terms of the differential entropy, it follows that

$$h(x) = H(X) + \lim_{\delta x \to 0} \log \delta x, \tag{1.17}$$

because the integration over $p(x)$ is one. In the limit as δx approaches zero, $\log \delta x$ approaches minus infinity. Hence, the differential entropy of a continuous random variable is negative infinitively large. However, this problem can be circumvented by adopting the term $\log \delta x$ as reference. This is common in case of comparing two differential entropies which have a common reference $(-\infty)$ and hence their relative entropy is positive.

When the Gaussian probability density is used as a common reference the relative entropy is called negentropy $J(X)$.

$$J(X) = D(p(x) \| p_G(x)) \tag{1.18}$$

$$= \int p(x) \log \frac{p(x)}{p_G(x)} dx$$

$$= \int p(x) \log p(x) dx - \int p(x) \log p_G(x) dx$$

$$= H_G(X) - H(X) \tag{1.19}$$

where $H_G(X)$ is the entropy of the Gaussian distribution with the same mean and variance as $p(x)$ and $H(X)$ is the entropy [2] of the random variable x. The integral $\int p(x) \log p_G(x) dx$ is the entropy of a Gaussian distribution for any distribution of $p(x)$ when $p(x)$ and $p_G(x)$ yield the same variance (Cover and Thomas, 1991, page 234).

1.3.2 Maximum Entropy

Under certain constraints it is possible to find a RV whose p.d.f. has the maximal entropy. The maximum entropy of a distribution is derived for (1) an amplitude bounded RV and for (2) a RV with fixed variance.

Theorem 1.1 (Maximum entropy of an amplitude bounded RV) *The entropy of an amplitude bounded RV is $H(X) \leq \log|N_X|$ where $|N_X|$ denotes the number of elements in the range of X, with equality if and only if X has a uniform distribution over N_X.*

Proof 1.1 *Let $q(x) = \frac{1}{|N_X|}$ be the uniform probability distribution function over N_X and let $p(x)$ be the probability distribution for X. Then*

$$\begin{aligned} D(p\|q) &= \sum p(x)\log\frac{p(x)}{q(x)} \\ &= \sum p(x)\log|N_X| - \sum p(x)\log\frac{1}{p(x)} \\ &= \log|N_X| - H(X). \end{aligned} \tag{1.20}$$

Since the relative entropy is non-negative it follows that

$$0 \leq D(p\|q) = \log|N_X| - H(X) \tag{1.21}$$

and therefore,

$$H(X) \leq \log|N_X|. \tag{1.22}$$

Hence, the uniform distribution has the highest entropy when X is of a given amplitude range.

Theorem 1.2 (Maximum entropy of a RV with fixed variance) *Let the continuous random variable x have zero mean and variance σ_x^2 then $H(x) \leq \frac{1}{2}\log(2\pi e)\sigma_x^2$ (entropy of x is always smaller than the entropy of a normal distribution) with equality iff $x \propto N(0,\sigma_x^2)$ where $N(0,\sigma_x^2)$ is the normal distribution.*

Proof 1.2 *Let $p(x)$ be any density satisfying $\int p(x)x^2dx = \sigma_x^2$. Let $N(0,\sigma_x^2)$ be a density of a normal distribution with zero-mean. Then*

$$\begin{aligned} 0 &\leq D(p\|N(0,\sigma_x^2)) \\ &\leq \int p(x)\log\frac{p(x)}{N(0,\sigma_x^2)} \\ &\leq -H(x) - \int p(x)\log N(0,\sigma_x^2) \\ &\leq -H(x) - \int N(0,\sigma_x^2)\log N(0,\sigma_x^2) \\ &\leq -H(x) + H(N(0,\sigma_x^2)) \\ H(x) &\leq H(N(0,\sigma_x^2)) \end{aligned} \tag{1.23}$$

where the substitution $\int p(x)\log N(0,\sigma_x^2) = \int N(0,\sigma_x^2)\log N(0,\sigma_x^2)$ follows from the fact that $p(x)$ and $N(0,\sigma_x^2)$ yield the same variance of the quadratic form $\log N(0,\sigma_x^2)$ (Cover and Thomas, 1991; page 234). The last inequality in eq.1.23 states that for a given variance, the normal distribution has the highest entropy.

1.4 ARTIFICIAL NEURAL NETWORKS

Artificial neural networks provide a general and practical method for learning functions from examples. The works have been inspired in part by the observation of biological learning systems. The task of understanding how the brain works is one of the outstanding unsolved problems in science (Churchland and Sejnowski, 1992). Neural network models are intended to elucidate how computation is is performed in neurons. The use of neural networks in engineering is to create machines that can learn. In particular, neural networks were used to perform pattern recognition such that it learns to recognize handwritten characters and spoken words. Another motivation arises from complex systems since neural networks can be build of very complex nets of interconnected neurons. An interesting property is the adaptive behavior of the complex system towards its environment.

Typically an artificial neural network can be characterized with three specifications: architecture, activation function and learning rule. The architecture specifies what variables are involved in the model and their topological relationships. The variables in a neural network are the weights of the connections between neurons and the activities of the neurons. The activation function and the weights describe the dynamics between the input and the output of a neuron. The learning rule specifies the way in which the neural network's weights change with time. The learning rule usually depends on the activities of the neurons, the weights and the input and output values of the neural network. If the learning algorithm depends on additional target values supplied by a teacher by labeling the data the learning process is called supervised learning. However, if the learning rules can be derived from objective functions and the data is not prelabeled the learning process is unsupervised. Alternatively, learning rules are created from heuristics.

The terminologies used for neural networks are explained in the following example. Figure 1.3 shows a simple single neuron that takes input values $x_1, x_2, \cdots, x_N$ and generates an output signal y. The architecture is a feedforward system because the connections are directed from the input to the output of the neuron. The activation function can be performed in two steps: First, the activation is computed as the inner product of a given n-dimensional input vector $\mathbf{x} = [x_1, x_2, \cdots, x_N]^T$ and a learnable n-dimensional weight vector $\mathbf{w} = [w_1, w_2, \cdots, w_N]^T$

$$u = \sum_{i}^{N} w_i x_i. \tag{1.24}$$

Second, the output y is computed as a function $f(u)$ of the activation. There are several possible activation functions, e.g.

- Linear: $y(u) = u$
- Sigmoid (logistic function): $y(u) = \frac{1}{1+\exp(-u)}$
- Threshold function:

$$y(u) = \begin{cases} 1 & : \quad u > 0 \\ -1 & : \quad u \leq 0 \end{cases}$$

An objective function, error function or cost function [3] can be defined as a function of $\mathbf{w}$ to measure how well the network solves a given task. The objective function is a sum of terms, one for each input-target pair x_k, T_k, measuring how close the scalar output $y(\mathbf{x}; \mathbf{w})$ is to the target. During training, the algorithm adjusts $\mathbf{w}$ in such a way as to find a $\mathbf{w}$ that minimizes the objective function. Instead of minimizing the function many learning rules make use of the gradient of the objective function with respect to $\mathbf{w}$.

The error at time t may be defined as the difference between the target response $T(t)$ and the actual response $y(t)$. A commonly used objective function is the mean-squared-error criterion

$$O_k(t) = E\{\frac{1}{2}\sum_k e_k^2(t)\}, \tag{1.25}$$

where k denotes the k-th neuron and the k-th error term is $e_k = y_k(t) - T_k(t)$. In the single neuron case the objective has only one term ($O(t) = \frac{1}{2}e^2(t)$). The expectation can be replaced by a training set of samples $\mathbf{x}$. Now, the online stochastic gradient descent learning rule can be derived by minimizing the gradient of the objective function.

$$\frac{\partial O(t)}{\partial w_i} = e(t)x_i(t). \tag{1.26}$$

The learning rule is then

$$\Delta w_i(t) = \epsilon \frac{\partial O}{\partial w_i} = \epsilon(T(t) - y(t))x_i(t) \tag{1.27}$$

where ϵ is the learning rate. This learning rule requires a teacher that labels the input vector with a target to specify what the neurons output should be.

1.4.1 Neural networks using unsupervised learning rules

Unsupervised learning rules are intended to learn from just a set of examples or observations $\mathbf{x}$. The purpose of an unsupervised or self-organizing learning rule is to discover significant patterns or features in the input data.

One important self-organizing principle is the Hebbian learning rule (Hebb, 1949). Roughly speaking, if there are two simultaneously active neurons on either side of a connection then the weight of that connection is increased. Hebbian learning can be demonstrated in a single neuron. The change in synaptic weight in a single neuron is

$$\Delta w_i(t) = \epsilon y(t)x_i(t). \tag{1.28}$$

The weight update is then

$$w_i(t+1) = w_i(t) + \epsilon y(t)x_i(t). \tag{1.29}$$

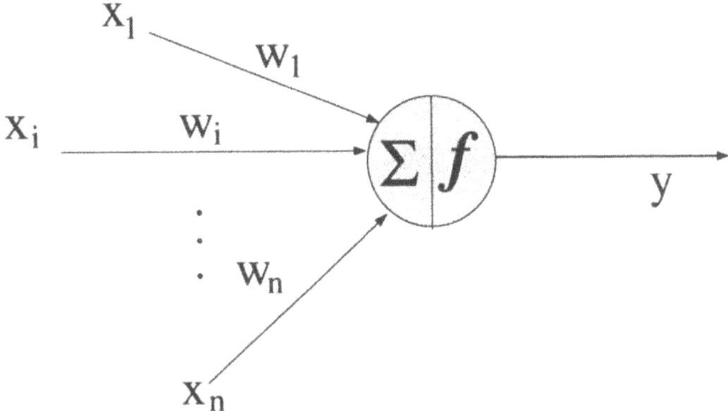

Figure 1.3. A simple single neuron consists of: inputs x_i, weights w_i and activation function $f(\mathbf{wx})$. The output is $y = f(\mathbf{wx})$.

However, this form of representation leads to an exponential growth of w_i with increasing training samples. One way to impose a limit in the growth of w_i is to incorporate a normalization term (Oja, 1982)

$$w_i(t+1) = \frac{w_i(t) + \epsilon y(t) x_i(t)}{(\sum_i (w_i(t) + \epsilon y(t) x_i(t))^2)^{1/2}}. \tag{1.30}$$

For small ϵ eq.1.30 can be approximately written as (Haykin, 1994b)

$$w_i(t+1) = w_i(t) + \epsilon y(t)[x_i(t) - y(t) w_i]. \tag{1.31}$$

The term $y(t)x_i(t)$ is the Hebbian self-amplification rule and $y(t)w_i$ is called the forgetting factor (Oja, 1982). The latter becomes more important with a stronger response $y(t)$. To demonstrate the function of Oja's learning rule 1000 data samples were randomly generated. The data were processed with the learning rule and $\mathbf{w}$ was updated at each sample with a fixed learning rate of $\epsilon = 0.005$. Figure 1.4 shows a 2-dimensional data space and the weight vector $\mathbf{w}$ that points in the direction of maximal variance. In fact, this process corresponds to finding a principal component. The mathematical proof of convergence and stability is in Hertz et al. (1991).

A common statistical method for analyzing data is Principal Component Analysis (PCA). It is closely related to Singular Value Decomposition (SVD) and in communication theory PCA is known as Karhunen-Loeve transform. PCA is a decorrelation-based method that finds a linear transformation $\mathbf{W}$ given the data $\mathbf{x}$ so that (1) the output vectors $\mathbf{u}$ are uncorrelated, (2) the basis vectors of $\mathbf{W}$ are orthogonal to each other and (3) the eigenvalues of $\mathbf{W}$ are ordered. To satisfy the decorrelation criterion (1), the covariance matrix output data $\mathbf{u}$ must be a diagonal matrix $\mathbf{D}$

$$\langle \mathbf{u}\mathbf{u}^T \rangle = \mathbf{D}, \tag{1.32}$$

where $\mathbf{u} = \mathbf{W}\mathbf{x}$ and the entries of the diagonal matrix $\mathbf{D}$ contain eigenvalues of the linear system $\mathbf{W}$. There are many matrices $\mathbf{W}$ that satisfy eq.1.32. The PCA

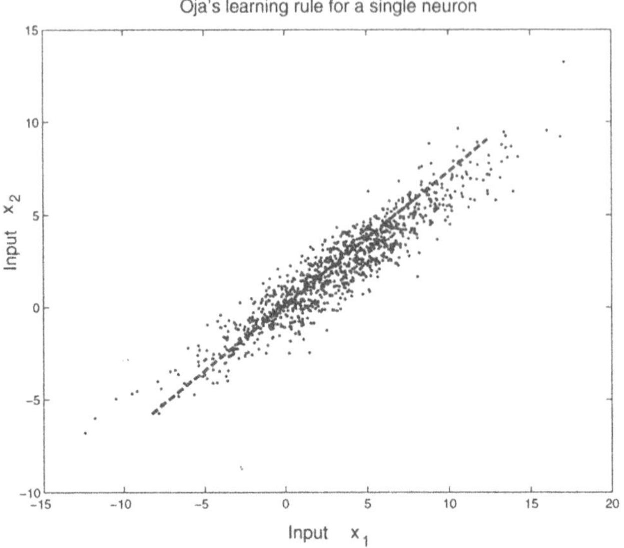

Figure 1.4. Oja's learning rule applied in a single neuron. The weight vector (doted line) finds the direction of maximum variance.

solution is uniquely defined and gives a matrix $\mathbf{W}_p$ such that the basis vectors of $\mathbf{W}_p$ are orthogonal to each other and ordered according to the eigenvalues. The orthogonality (2) is satisfied when

$$\mathbf{W}_P \mathbf{W}_P^T = \mathbf{S}, \tag{1.33}$$

where $\mathbf{S}$ is a diagonal scaling matrix. This solution is also called the global decorrelating solution because the PCA filters (rows of $\mathbf{W}_P$) are ordered (3) according to the amplitude spectrum of the data.

It has been demonstrated how a single neuron is capable of finding the direction of maximal variance. This function can be extended in a feedforward single layer of neurons, that can perform PCA. Consider a neural network shown in figure 1.5. The network has N inputs and N neurons. Each neuron in the output layer of the network is linear. The synaptic weights w_{ij} are connecting the i-th inputs and the j-th neuron, where here $i = 1, \cdots, N$ and $j = 1, \cdots, N$. The output of neuron j at time t is produced in response to the set of inputs $x_i(t)$

$$y_j(t) = \sum_{i=1}^{N} w_{ij}(t) x_i(t). \tag{1.34}$$

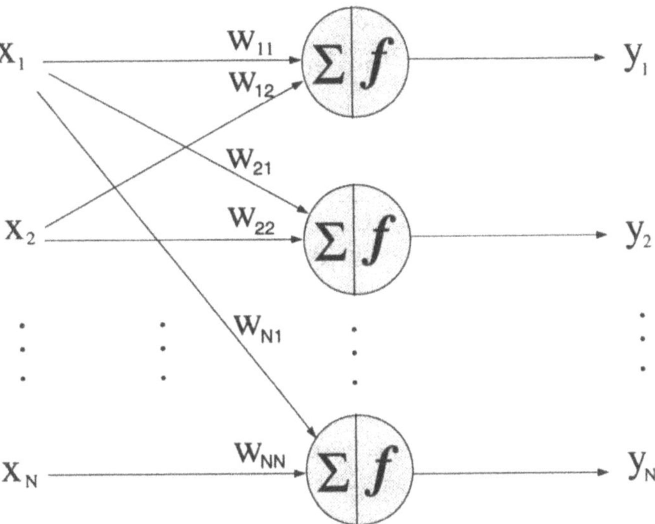

Figure 1.5. Single layer feedforward neural network. The input vector is $\mathbf{x}$, the synaptic weight matrix is $\mathbf{W}$ and the output vector is $\mathbf{y}$.

The weight $w_{ij}(t)$ is adapted in the generalized form of Hebbian learning (Sanger, 1989)

$$\Delta w_{ij}(t) = \epsilon \left[y_j(t)x_i(t) - y_j(t) \sum_{k=1}^{i} w_{kj}(t) y_k(t) \right] . \tag{1.35}$$

Eq.1.35 is known as the Generalized Hebbian Algorithm (GHA) and can be written in the matrix notation as follows

$$\Delta \mathbf{w}_j(t+1) = \epsilon y_j(t)\mathbf{x}'(t) - \epsilon y_j^2(t)\mathbf{w}_j(t), \tag{1.36}$$

where

$$\mathbf{x}'(t) = \mathbf{x}(t) - \sum_{k=1}^{j} \mathbf{w}_k(t) y_k(t). \tag{1.37}$$

Based on the number of neurons in the GHA network the following observations are made

- For $j = 1$ and $\mathbf{x}' = \mathbf{x}$, eq.1.36 reduces to eq.1.31 for a single neuron. This neuron will discover the first principal component, i.e. the largest eigenvalue and associated eigenvector of the input vector $\mathbf{x}(t)$.
- For $j = 2$ and $\mathbf{x}' = \mathbf{x} - \mathbf{w}_1(n)y_1(t)$, if the first neuron has learned the first principal component, the second neuron sees $\mathbf{x}'$ from which the first eigenvector of

the correlation matrix has been removed. Therefore, the second neuron extracts the first principal component of $\mathbf{x}'$ which is equivalent to the second principal component, i.e. the second largest eigenvalue of the data. It is perpendicular to the first principal component.

This can be proceeded for the remaining neurons and it is clear that each output represents a particular eigenvector of the correlation matrix of the input vector, and that the individual outputs are ordered by decreasing eigenvalue. The convergence theorem of the GHA is shown in Haykin (1994b).

There are several applications associated with PCA. The PCA transformation is designed in such a way that the data set may be represented by a number of reduced number of effective features and still retain most of the intrinsic information content of the data. It therefore performs a dimensionality reduction and may be used for example in image coding where an image is reconstructed given only the first few principal components. This data compression technique allows a transmission of reduced data through a limited-bandwidth channel.

1.4.2 The Principle of Maximum Entropy Preservation

The principle of maximum entropy preservation (infomax) is closely related to the concept of channel capacity which is Shannon's (1948) second theorem of the mathematical theory of communication: the channel coding theorem. A neural network may be viewed as a communication system receiving inputs and efficiently coding it.

The relation of sensory coding strategy and neural coding was pointed out by Barlow (1961). He proposed Redúndancy reduction as a property for the study of human sensory coding. The idea was to formulate an objective function based on information theoretic criteria that one thinks a neural code should satisfy. In this context Barlow suggested that neurons in the receptive field such as the visual reception and the olfactory reception reduce redundant information that leads to factorial code. This assumes that each neuron is independent of the features encoded by the other neurons. Atick (1992) proposed a linear neural network for visual processing to perform redundancy reduction. The method is similar to Linsker's infomax principle that was formulated in a linear neural network. Linsker's (1989) principle of maximum information preservation suggested the following:

> The transformation of a vector $\mathbf{x}$ observed in the input layer of a neural network to a vector $\mathbf{y}$ produced in the output of the output layer jointly maximize information about the activities in the input layer. The parameter to be maximized is the average mutual information between the input vector $\mathbf{x}$ and the output vector y, in the presence of processing noise.

Here, the goal of the neurons in the network is to maximize the mutual information between the sensory inputs and the network outputs. This maximization can be formulated as the principle of maximum entropy preservation in a single layer feedforward network under certain circumstances which are dependent on the neural output distributions, the type of activation function and the noise model. For simplicity, the application of the infomax principle is demonstrated in a network with two neurons as shown in figure 1.6. It is assumed that the neurons include additive

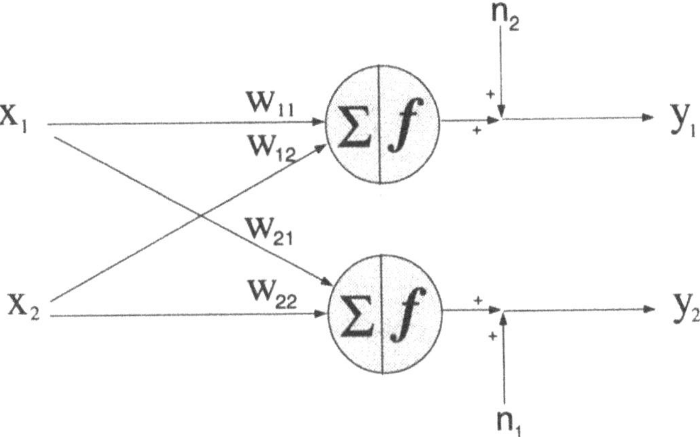

Figure 1.6. A linear neural network with two inputs and two neurons. Gaussian noise sources n_1 and n_2 are added at the neurons.

noise sources n_1 and n_2 and a linear activation function $y = f(u) = u$. In the following, the mutual information in the network is computed as a function of the noise variance of n_1 and n_2. The outputs of the neurons in figure 1.6 are

$$\begin{aligned} y_1 &= w_{11}x_1 + w_{12}x_2 + n_1 \\ y_2 &= w_{21}x_1 + w_{22}x_2 + n_2. \end{aligned} \tag{1.38}$$

The mutual information between $\mathbf{x}$ and $\mathbf{y}$ is

$$I(\mathbf{x}; \mathbf{y}) = H(\mathbf{x}) + H(\mathbf{y}) - H(\mathbf{x}, \mathbf{y}). \tag{1.39}$$

The joint entropy $H(\mathbf{x}, \mathbf{y}) = H(\mathbf{x}) - H(\mathbf{n})$ because the linear mapping from $\mathbf{x}$ to $\mathbf{y}$ is determined by the weight matrix $\mathbf{W}$ and the noise sources. It is therefore deterministic up to the noise sources. Another interpretation is that $I(\mathbf{x}; \mathbf{y}) = H(\mathbf{y}) - H(\mathbf{y}|\mathbf{x})$ and the conditional entropy $H(\mathbf{y}|\mathbf{x})$ is the information that the neuron conveys about $\mathbf{n}$ rather then about $\mathbf{x}$. Therefore, $H(\mathbf{y}|\mathbf{x}) = H(\mathbf{n})$ and it follows that

$$I(\mathbf{x}; \mathbf{y}) = H(\mathbf{y}) - H(\mathbf{n}). \tag{1.40}$$

Information maximization in the linear network is achieved when y_1 and y_2 have Gaussian distributions. Note that the entropy of a distribution with a fixed variance is maximum for the Gaussian distribution (see section 1.3.2). This is here the case since the amplitudes of y_1 and y_2 are not bounded and the activation functions are linear. The entropy terms in eq.1.40 can be computed as follows: A n-dimensional multivariate Gaussian distribution is defined as

$$p(\mathbf{y}) = \frac{1}{\sqrt{(2\pi)^N \det(\mathbf{C})}} \exp\left(-\frac{1}{2}\mathbf{y}^T\mathbf{C}^{-1}, \mathbf{y}\right) \tag{1.41}$$

where $\det(\mathbf{C})$ is the determinant of the covariance matrix $\mathbf{C} = \langle \mathbf{y}\mathbf{y}^T \rangle$. The entropy of the output neurons $\mathbf{y}$ is then (Cover and Thomas, 1991)

$$H(\mathbf{y}) = \frac{1}{2} \log \left[(2\pi e)^N \det(\mathbf{C}) \right]. \tag{1.42}$$

The joint entropy for a Gaussian distributed noise signal with variance σ_n for $N = 2$ is (Haykin, 1994b)

$$H(n_1, n_2) = H(n_1) + H(n_2) = \log(2\pi e \sigma_n^2), \tag{1.43}$$

where it is assumed that both noise variances are equivalent. The mutual information from eq.1.40, eq.1.42 and eq.1.43 for the network is then

$$I(\mathbf{x}; \mathbf{y}) = \log \left(\frac{\det(\mathbf{C})}{\sigma_n^2} \right). \tag{1.44}$$

The mutual information between the output vector $\mathbf{y}$ and the input vector $\mathbf{x}$ in eq.1.44 depends on the noise variance. Maximizing eq.1.44 is equivalent to maximizing the determinant of the covariance matrix $\mathbf{C}$. The covariance matrix $\mathbf{C}$ in the network with two neurons is

$$\begin{aligned} \mathbf{C} &= E\{\mathbf{y}\mathbf{y}^T\} = \langle \mathbf{y}\mathbf{y}^T \rangle \\ &= \begin{bmatrix} c_{11} & c_{12} \\ c_{21} & c_{22} \end{bmatrix}. \end{aligned} \tag{1.45}$$

The elements of the covariance matrix can be written as variances of y_1 and y_2

$$\begin{aligned} c_{11} &= \sigma_1^2 + \sigma_n^2 \\ c_{21} &= c_{12} = \sigma_1 \sigma_2 \rho_{21} \\ c_{22} &= \sigma_2^2 + \sigma_n^2, \end{aligned} \tag{1.46}$$

where ρ_{21} is the correlation coefficient of y_2 and y_1. The determinant of the covariance matrix is

$$\begin{aligned} \det(\mathbf{C}) &= c_{11}c_{22} - c_{12}c_{21} \\ &= \sigma_n^4 + \sigma_n^2(\sigma_1^2 + \sigma_2^2) + \sigma_1^2\sigma_2^2(1 - \rho_{21}^2). \end{aligned} \tag{1.47}$$

Eq.1.47 can be maximize under two different noise-level conditions (Haykin, 1994b)

- **Large noise variance**: The third term of eq.1.47 can now be neglected. For a fixed noise variance maximizing $\det(\mathbf{C})$ is achieved by maximizing the noise terms either σ_1^2 or σ_2^2. Linsker concludes that a high noise level favors redundancy at the outputs, i.e. the two neurons compute the same linear combination of inputs, given that there is only one such combination that yields a response with maximum variance.

- **Low noise variance**: The third term in eq.1.47 becomes relatively important and the mutual information is maximized by an optimal tradeoff between two options:

keeping the output variance σ_1^2 and σ_2^2 large, and making the outputs y_1 and y_2 uncorrelated. Here, the neurons try to compute different linear combinations of inputs, even though such a choice may result in a reduced output variance.

In this example, the mutual information was maximized between the inputs and the outputs of a neural network. In the case of low additive noise signals the infomax principle will decorrelate the output signals. Note that this derivation was based on the assumptions that the activation functions were linear. The assumptions about the network properties are important. The chapter on ICA shows how the infomax principle can be used to perform ICA by modifying these assumptions.

1.5 HIGHER-ORDER STATISTICS

This section presents the basic definitions and relations of moments and cumulants which are important properties describing higher-order statistics of a RV.

Higher-order statistics of a RV are necessary to describe random process in which the behavior of the RV is non-Gaussian. A Gaussian process can be described entirely using second-order statistics, i.e. mean and variance. However, non-Gaussian processes have in addition to first- and second-order statistics higher-order information that describe the form of the process (e.g. skewness, kurtosis). Moments and cumulants can be used to describe any distribution of a RV. For simplicity, moments and cumulants are described using a single RV whereas cross-cumulants involve at least two RVs. Books and articles on higher-order statistics are DeGroot (1986); Stuart and Ord (1987); Papoulis (1990); Cadzow (1996).

1.5.1 Moments

A stationary random time series can be thought of as being generated by a sequence of samples of an underlying generating random variable. The p.d.f. is a continuous function of x and governs the generation of x. For any random variable, the p.d.f. of x can be described in terms of a set of discrete parameters called moments

$$\mu_x(n) = E\{x^n\} = \int_{-\infty}^{+\infty} x^n p(x) dx \tag{1.48}$$

where $\mu_x(n)$ denotes the nth-order moment. The first-order moment ($\mu_x(1) = m_x$) is referred to as the mean value of x, i.e. it corresponds to the center of the distribution. Central moments provide a set of parameters that describe the manner in which the distribution is about its mean value m_x.

$$m_x(n) = E\{(x - m_x)^n\}. \tag{1.49}$$

Central moments m and moments μ are identical when the mean is zero. The second-order central moment is referred to as the variance: $\sigma_x^2 = m_x^2$. The third-order central moment measures the skewness of the p.d.f. about its mean value. It is zero for a symmetrical p.d.f.s. The fourth-order central moment is used to measure the excess or flatness (i.e. kurtosis) of the p.d.f.

Consider now some properties of the Fourier transform of the random variable x i.e. the expected value of the exponential function x which is commonly used in the signal processing community. The Fourier transform of the p.d.f. is called $\psi_x(\tau)$, the moment generating function (m.g.f.) of x

$$\psi_x(\tau) = \int_{-\infty}^{+\infty} e^{j\tau x} p(x) dx = E\{e^{j\tau x}\}. \tag{1.50}$$

It possesses all the properties associated with the Fourier transform.

Suppose that the m.g.f. of x exists for all values of τ in some interval around the point $\tau = 0$. It can be shown that the derivative $\partial\psi_x(\tau)/\partial\tau$ in eq.1.50 is equal to the expectation of the derivative

$$\frac{\partial\psi_x(\tau)}{\partial\tau} = \frac{\partial}{\partial\tau} E\{\exp(j\tau x)\}|_{\tau=0} = E\{\frac{\partial}{\partial\tau}\exp(j\tau x)|_{\tau=0}\}. \tag{1.51}$$

But since

$$\frac{\partial}{\partial\tau}\exp(j\tau x)|_{\tau=0} = x\exp(j\tau x)|_{\tau=0} = x, \tag{1.52}$$

it follows that

$$\frac{\partial\psi_x}{\partial\tau}|_{\tau=0} = E\{x\}. \tag{1.53}$$

The derivative of the m.g.f. at $\tau = 0$ is the mean of x. More generally, if the m.g.f. exists for all values of τ in an interval around the point $\tau = 0$, then it can be shown that the n-th derivative will satisfy

$$\psi_x^n(\tau = 0) = E\{x^n\}. \tag{1.54}$$

One important property of m.g.f. is that the sum of independent random variables has a simple form.

Theorem 1.3 (Property of m.g.f.) *$x_1, \cdots, x_n$ are independent random variables and $\psi_{(x_i)}(\tau)$ denote the corresponding m.g.f. for $i = 1, \cdots, n$. $\mathbf{x}$ is the sum of independent variables $\mathbf{x} = x_1 + \cdots + x_n$ the $\psi_{(\mathbf{x})}(\tau)$ denotes the m.g.f. of $\mathbf{x}$. Then for any value of τ it follows*

$$\psi_{\mathbf{x}}(\tau) = \prod_{i=1}^{n} \psi_{(x_i)}(\tau). \tag{1.55}$$

Proof 1.3 *By definition*

$$\psi_{\mathbf{x}}(\tau) = E\{\exp(\tau\mathbf{x})\} = E\{\exp(\tau(x_1 + \cdots + x_n))\} = E\{\prod_{i=1}^{n}\exp(\tau x_i)\}. \tag{1.56}$$

Since the random variables are independent the product of the exponentials is the product of the expected value of the exponentials. Hence,

$$\psi_{\mathbf{x}}(\tau) = \prod_{i=1}^{n} \psi_{x_i}(\tau). \tag{1.57}$$

1.5.2 Cumulants

The logarithm of the m.g.f. $\psi_x(\tau)$ is called the cumulant generating function (c.g.f.)

$$\Phi_x(\tau) = \log[E\{e^{j\tau x}\}]. \tag{1.58}$$

An interpretation of the cumulant generating function is obtained by making a Taylor series expansion in τ^k about its origin. The coefficients of the Taylor series term τ^k, multiplied by $(-j)^k$ is called the k-th order cumulant and is given by

$$c_k = (-j)^k \frac{\partial d^k \Phi_x(\tau)}{\partial d\tau^k}|_{\tau=0}. \tag{1.59}$$

Cumulants can characterize random variables as functions of mean and moments. For example, the first four cumulants for a single RV are (Stuart and Ord, 1987)

$$\begin{aligned} c_1 &= m_1 = \mu \\ c_2 &= m_2 = \sigma^2 \\ c_3 &= m_3 \\ c_4 &= m_4 - 3m_2^2, \end{aligned} \tag{1.60}$$

where m_n are central moments.

- The first-order cumulant is exactly the mean of x as derived in eq.1.53.
- The second-order cumulant is the variance of x.
- The third-order cumulant is the same the third-order moment.
- The fourth-order cumulant has the fourth-order moment and six other second-order moments which are summarized in $3m_2^2$ because there is only one RV.

1.5.3 Cross-cumulants

The term cross-cumulants describes cumulants of more than one RV. For example, the second-order cumulant of two RVs is defined as

$$c_2(x_1, x_2) = E\{m_1(x_1)m_1(x_2)\} \tag{1.61}$$

which are elements of the covariance matrix of two RVs. The 4th-order order cross-cumulants are defined as

$$\begin{aligned} c_4(x_1, x_2, x_3, x_4) = \; & E\{m_1(x_1)m_1(x_2)m_1(x_3)m_1(x_4)\} - \\ & E\{m_1(x_1)m_1(x_2)\}E\{m_1(x_3)m_1(x_4)\} - \\ & E\{m_1(x_1)m_1(x_3)\}E\{m_1(x_2)m_1(x_4)\} - \\ & E\{m_1(x_1)m_1(x_4)\}E\{m_1(x_2)m_1(x_3)\}. \end{aligned} \tag{1.62}$$

Eq.1.62 reduces to the 4th-order cumulants (kurtosis) for a single variable in eq.1.60 when $x_1 = x_2 = x_3 = x_4$.

1.6 SUMMARY

This chapter briefly summarized the basics of the theories that are necessary to understand the material presented in the following theory chapters.

Bayesian probability theory provides a general framework for estimation and inference. Information theory gives a measure of redundancy which is related to channel capacity for the transmission rate in communications. Maximum entropy is achieved for a signal with given variance when its p.d.f. is Gaussian. For an amplitude bounded signal, the entropy is maximum when its distribution is uniform. Artificial neural networks are biologically inspired and can be used in a broad and interdisplinary field. The focus in this chapter was on unsupervised learning algorithms for simple feedforward neural networks. A single neuron can find the direction of maximum variance using a Hebbian learning rule and the extension to a single layer neural network is capable of performing principle component analysis. Linsker (1989) suggested that the objective of a neural network is to maximize the mutual information between the inputs and outputs. A discussion on a simple two by two neural network with varying noise quantities showed that in the low noise case maximizing the entropy at the outputs reduced the redundancy among the outputs for an optimal information flow. Moments and cumulants are parameters that convey information about the distribution of a random variable. In particular, non-Gaussian variables may be described using higher-order moments and cumulants.

The next chapter shows how learning algorithms can be derived based on the presented theories that perform blind source separation.

Notes

1. It is available online via his homepage: `http://wol.ra.phy.cam.ac.uk/mackay/itprnn/#book`

2. Note that throughout the book, $H(.)$ is referred to as differential entropy. $h(x)$ was merely used in this section to distinguish between the entropies of a discrete random variable (RV) and a continuous RV.

3. The three terms can be used interchangeably in this context.

2 INDEPENDENT COMPONENT ANALYSIS

The world beyond second-order statistics
Anthony Bell

2.1 OVERVIEW

The goal of blind source separation (BSS) is to recover independent sources given only sensor observations that are linear mixtures of independent source signals. The term *blind* indicates that both the source signals and the way the signals were mixed are unknown. Independent Component Analysis (ICA) is a method for solving the blind source separation problem. It is a way to find a linear coordinate system (the unmixing system) such that the resulting signals are as statistically independent from each other as possible. In contrast to correlation-based transformations such as Principal Component Analysis (PCA), ICA not only decorrelates the signals (2nd-order statistics) but also reduces higher-order statistical dependencies.

Two different research communities have considered the analysis of independent components. On one hand, the study of separating mixed sources observed in an array of sensors has been a classical and difficult signal processing problem. The seminal work on blind source separation was by Herault and Jutten (1986) where they introduced an adaptive algorithm in a simple feedback architecture that was able to separate several unknown independent sources. Their approach has been further developed by Jutten and Herault (1991), Karhunen and Joutsensalo (1994), and

Cichocki et al. (1994). Comon (1994) elaborated the concept of independent component analysis and proposed cost functions related to the approximate minimization of mutual information between the sensors.

In parallel to blind source separation studies, unsupervised learning rules based on information theory were proposed by Linsker (1992). The goal was to maximize the mutual information between the inputs and outputs of a neural network. This approach is related to the principle of redundancy reduction suggested by Barlow (1961) as a coding strategy in neurons. Each neuron should encode features that are as statistically independent as possible from other neurons over a natural ensemble of inputs; decorrelation as a strategy for visual processing was explored by Atick (1992). Nadal and Parga (1994) showed that in the low-noise case, the maximum of the mutual information between the input and output of a neural network implied that the output distribution was factorial; that is, the multivariate probability density function (p.d.f.) can be factorized as a product of marginal p.d.f.s. Roth and Baram (1996) and Bell and Sejnowski (1995) independently derived stochastic gradient learning rules for this maximization and applied them, respectively, to forecasting, time series analysis, and the blind separation of sources. Bell and Sejnowski (1995) put the blind source separation problem into an information-theoretic framework and demonstrated the separation and deconvolution of mixed sources. Their adaptive methods are more plausible from a neural processing perspective than the cumulant-based cost functions proposed Comon (1994). A similar adaptive method for source separation was proposed by Cardoso and Laheld (1996).

The original infomax learning rule for blind separation by Bell and Sejnowski (1995) was suitable for super-Gaussian sources, i.e. sources with probability density functions (p.d.f.s) sharply peaked with heavy tails and positive kurtosis (normalized 4th-order cumulant). As illustrated in Bell and Sejnowski (1995) their algorithm fails to separate sources that have negative kurtosis (sub-Gaussian). An extension of the infomax algorithm of Bell and Sejnowski (1995) is presented in Lee et al. (1998b) that is able to blindly separate mixed signals with sub- and super-Gaussian source distributions. This was achieved by using a simple type of learning rule first derived by Girolami (1997b) by choosing negentropy as a projection pursuit index. Parameterized probability distributions that have sub- and super-Gaussian regimes were used to derive a general learning rule that preserves the simple architecture proposed by Bell and Sejnowski (1995), is optimized using the natural gradient by Amari (1998), and uses the stability analysis of Cardoso and Laheld (1996) to switch between sub- and super-Gaussian regimes.

There are two important properties in ICA: the natural gradient and the robustness in ICA against parameter mismatch. The natural gradient (Amari, 1998), or equivalently the relative gradient Cardoso and Laheld (1996) gives fast convergence. A simple nonlinearity used in the ICA learning rule is related to the source density model. However, it is robust against a parametric mismatch between the infomax density estimation and the true source density. Conditions are shown under which the infomax algorithm still converges to an ICA solution. Computer simulations demonstrate that the extended infomax algorithm can successfully separate 20 mixtures of the following sources: 10 sound tracks obtained from Pearlmutter, 6 speech

& sound signals used in (Bell and Sejnowski, 1995), 3 uniformly distributed sub-Gaussian noise signals and one noise source with a Gaussian distribution. However, instantaneous mixing and unmixing simulations are *toy* problems and the real challenge lies in dealing with real-world data. Results on biomedical recordings are shown in chapter 6.

This chapter has two intentions: First, it serves as an introduction to ICA by formulating the blind source separation problem, by comparing the ICA method to PCA and by summarizing the Bell and Sejnowski (1995) infomax learning algorithm that separates super-Gaussian sources. Second, this chapter presents an extension of the infomax algorithm that is able to blindly separate mixed signals with sub- and super-Gaussian source distributions.

The organization of this chapter is as follows: Section 2.2 states the problem and the assumptions for ICA. Section 2.3 demonstrates why decorrelation-based algorithms fail to separate independent sources. Section 2.4 reviews the infomax approach by Bell and Sejnowski (1995) and in section 2.5 the extended infomax learning algorithm is presented that can separate mixtures of sub- and super-Gaussian sources. This learning algorithm is applied to simulations in section 2.7. The results of the extended infomax algorithm are compared to the original infomax learning algorithm. Sections 2.8.1 gives an intuitive explanation for the 'natural' gradient proposed by Amari et al. (1996); Amari (1998) or 'relative' gradient proposed by Cardoso and Laheld (1996). Section 2.8.2 gives an intuitive explanation for the robustness in blind source separation. Finally, several ICA issues are discussed in section 2.9.

2.2 PROBLEM STATEMENT AND ASSUMPTIONS

Assume that there is an M-dimensional zero-mean vector $\mathbf{s}(t) = [s_1(t), \cdots, s_M(t)]^T$, such that the components $s_i(t)$ are mutually independent. The vector $\mathbf{s}(t)$ corresponds to M independent scalar-valued source signals $s_i(t)$. The multivariate p.d.f. of the vector can be rewritten as the product of marginal independent distributions.

$$p(\mathbf{s}) = \prod_{i=1}^{M} p_i(s_i). \tag{2.1}$$

A data vector $\mathbf{x}(t) = [x_1(t), \cdots, x_N(t)]^T$ is observed at each time point t, such that

$$\mathbf{x}(t) = \mathbf{A}\mathbf{s}(t), \tag{2.2}$$

where A is a full rank $N \times M$ scalar matrix. As the components of the observed vectors are no longer independent, the multivariate p.d.f. will not satisfy the p.d.f. product equality. If the components of $\mathbf{s}(t)$ are such that at most one source is normally distributed then it is possible to extract the sources $\mathbf{s}(t)$ from the received mixtures $\mathbf{x}(t)$ (Comon, 1994). The mutual information of the observed vector is given by the Kullback-Leibler (KL) divergence of the multivariate density from the

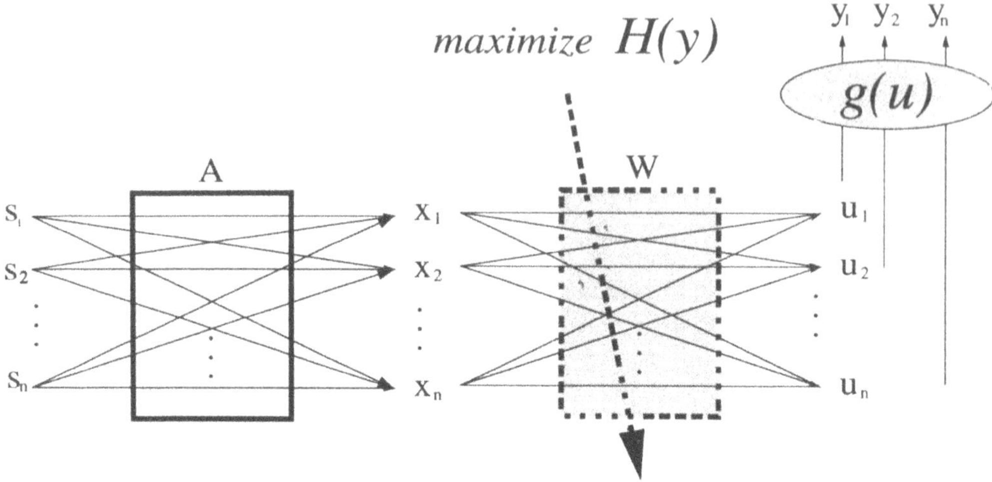

Figure 2.1. The instantaneous mixing and unmixing model. Independent sources $\mathbf{s}$ become mixed by $\mathbf{A}$. The observed sources are $\mathbf{x}$. The goal is to learn $\mathbf{W}$ that inverts the mixing $\mathbf{A}$ and $\mathbf{u}$ are the estimates of the recovered sources. The infomax approach is one way to find the unmixing system $\mathbf{W}$. It requires a nonlinear transfer function $g(\mathbf{u})$.

product of the marginal (univariate) densities

$$I(x_1, x_2, \cdots, x_N) = \int_{-\infty}^{+\infty} \int_{-\infty}^{+\infty} \cdots \int_{-\infty}^{+\infty} p(x_1, x_2, \cdots, x_N) \log \frac{p(x_1, x_2, \cdots, x_N)}{\prod_{i=1}^{N} p_i(x_i)} dx_1 dx_2 \cdots dx_N. \quad (2.3)$$

For simplicity,

$$I(\mathbf{x}) = \int p(\mathbf{x}) \log \frac{p(\mathbf{x})}{\prod_{i=1}^{N} p_i(x_i)} d\mathbf{x}. \quad (2.4)$$

The mutual information will always be positive and will only equal zero when the components are independent (Cover and Thomas, 1991).

The goal of ICA is to find a linear transformation $\mathbf{W}$ of the dependent sensor signals $\mathbf{x}$ that makes the outputs as independent as possible

$$\mathbf{u}(t) = \mathbf{W}\mathbf{x}(t) = \mathbf{W}\mathbf{A}\mathbf{s}(t), \quad (2.5)$$

where $\mathbf{u}$ is an estimate of the sources. The sources are exactly recovered when $\mathbf{W}$ is the inverse of $\mathbf{A}$ up to a permutation and scale change.

$$\mathbf{P} = \mathbf{R}\mathbf{S} = \mathbf{W}\mathbf{A}, \quad (2.6)$$

where $\mathbf{R}$ is a permutation matrix and $\mathbf{S}$ is the scaling matrix. The two matrices define the performance matrix $\mathbf{P}$ so that if $\mathbf{P}$ is normalized and reordered a perfect separation leads to the identity matrix. For the linear mixing and unmixing model, he following assumptions are adopted (Comon, 1994; Cardoso and Laheld, 1996)

1. The number of sensors is greater than or equal to the number of sources $N \geq M$.
2. The sources $\mathbf{s}(t)$ are at each time instant mutually independent.
3. At most one source is normally distributed.
4. No sensor noise or only low additive noise signals are permitted.

Assumption 1 is needed to make $\mathbf{A}$ a full rank matrix. Assumption 2 is the basis of ICA and can be expressed as follows

$$p(\mathbf{s}(t)) = \prod_{i=1}^{M} p(s_i(t)). \tag{2.7}$$

For assumption 3 the unmixing of two Gaussian sources is ill posed when the sources are white random processes. Non-white Gaussian processes may be recovered with time-decorrelation methods if they have different spectra (Molgedey and Schuster, 1994). However, pure Gaussian processes are rare in real data. Assumption 4 is necessary to satisfy the infomax condition, in which the mutual information between outputs is only minimized for the low noise case (Linsker, 1992; Nadal and Parga, 1994). However, one can imagine that noise is an independent source itself and if as many sensor outputs are available as the number of sources the noise signal can be segregated from the mixtures.

2.3 THE POVERTY OF PCA

Principal Component Analysis (PCA) is a popular tool for multivariate data analysis. Chapter 1, showed a single layer feedforward neural network that can find the principal components in the data using the Generalized Hebbian Algorithm (GHA). PCA is a decorrelation-based method and Linsker (1989) showed that performing infomax in a linear neural network decorrelated the outputs in the low noise case. Here, simple examples are presented to illustrate that decorrelation-based algorithms such as PCA cannot be used to separate independent sources. The reason for comparing PCA with ICA is due to the familiarity of PCA to most readers and due to some close relationships between ICA and PCA.

There are many ways to perform decorrelation. As described in chapter 1, PCA filters (rows of $\mathbf{W}_P$) give an orthogonal solution. Another decorrelation method is the symmetrical solution which assumes that the filters of $\mathbf{W}_Z$ are symmetrical (zero-phase) and therefore they are called Zero-phase Component Analysis (ZCA) (Bell and Sejnowski, 1997). If $\mathbf{W}_Z$ is symmetrical then $\mathbf{W}_Z = \mathbf{W}_Z^T$. PCA can be computed using the GHA but a simple singular value decomposition method (Jolliffe, 1986; Kaliath, 1980) can be used as well

$$\mathbf{W}_P = \mathbf{D}^{-\frac{1}{2}}\mathbf{E}^T, \tag{2.8}$$

where $\mathbf{D}$ is the diagonal matrix of eigenvalues and $\mathbf{E}$ are the eigenvectors of the covariance matrix (columns of $\mathbf{E}$) so that $\mathbf{D}$ and $\mathbf{E}$ satisfy

$$\mathbf{EDE}^{-1} = \langle \mathbf{xx}^T \rangle. \tag{2.9}$$

PCA is now applied to two simple blind source separation problems. The first simulation example involves two uniformly distributed sources s_1 and s_2. The sources are linearly independent because the values of one source does not convey any information about the other source. Figure 2.2 (a) shows the scatter-plot of the two original sources. The sources are linearly mixed as follows

$$\begin{aligned} \mathbf{x} &= \mathbf{As} \\ \begin{bmatrix} x_1 \\ x_2 \end{bmatrix} &= \begin{bmatrix} 1 & 2 \\ 1 & 1 \end{bmatrix} \begin{bmatrix} s_1 \\ s_2 \end{bmatrix}. \end{aligned} \tag{2.10}$$

Figure 2.2 (b) shows the scatter-plot of the mixtures. The distribution along the axis x_1 and x_2 are now dependent and the form of the density is stretched according to the mixing matrix. Applying PCA to the mixed data x_1 and x_2 results in two principal components. The first principal component is the axis accounting for the highest variance in the data and the second principal component is the axis orthogonal to first principal component axis. Figure 2.2 (c) shows the PCA solution and the result differs from the original since the two principal axis are still dependent. Due to the decorrelation property of PCA the data is sphered. However, the data needs to be rotated to correspond to the original source solution. The ICA solution in figure 2.2 (d) does not only sphere (decorrelate) the data but also effectively rotates it such that the axis of u_1 and u_2 have the same direction as the axis of s_1 and s_2. Since PCA is a decorrelation-based method and its objective is to decorrelate signals and not to make them independent this result may not be surprising. Independence however, and hence the separation of independent sources is achieved when the joint p.d.f. factorizes. When the sources are Gaussian distributed the joint p.d.f. for a multivariate Gaussian distribution factorizes when the marginal p.d.f.s are decorrelated because a Gaussian process is entirely described by 1st- and 2nd-order statistics. When the sources are not Gaussian distributed the joint p.d.f. will not be factorized when the sources are decorrelated. As described in section 1.5 non-Gaussian distributions can be parameterized using higher-order moments and cumulants. Therefore, the joint p.d.f. of non-Gaussian densities will only factorize, i.e. achieve independence when the sources are decorrelated and their higher-order correlations are removed as well.

The second simulation example in figure 2.3 (a) and (b) shows the time course of two speech signals s_1 and s_2. The signals are linearly mixed as in the previous example in eq.2.10.

Figure 2.3 (top) shows the two original sources, the linearly mixed signals x_1 and x_2 (second row), the recovered signals using PCA (third row) and the recovered signals using ICA (bottom). The performance matrix $\mathbf{P} = \mathbf{WA}$ is a measure for the separation quality. Due to possible scaling and permutation the performance matrix should be a diagonal matrix after normalizing and reordering. The performance

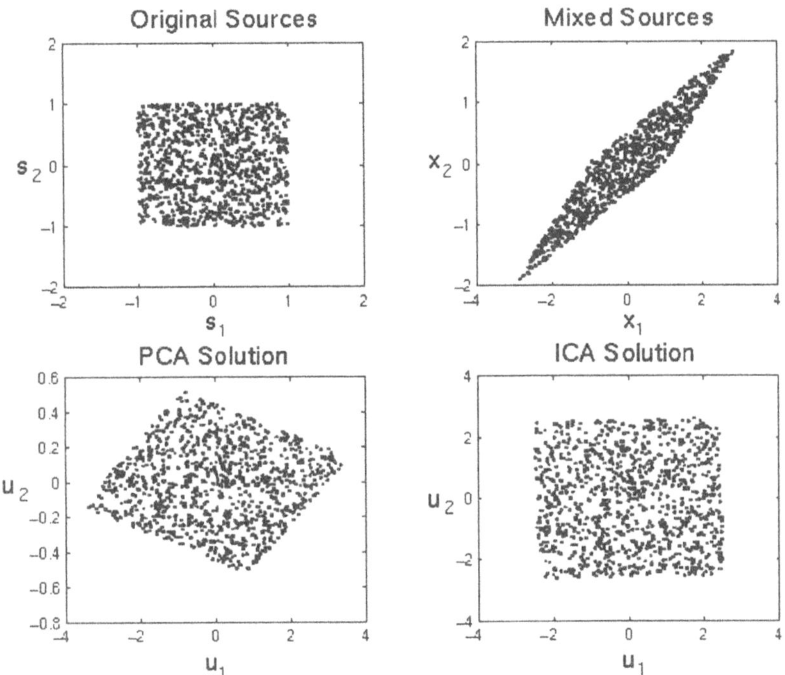

Figure 2.2. Top-left: scatter-plot of the original sources, Top-right: the mixtures, Bottom-left: the recovered sources using PCA, Bottom-right: the recovered sources using ICA.

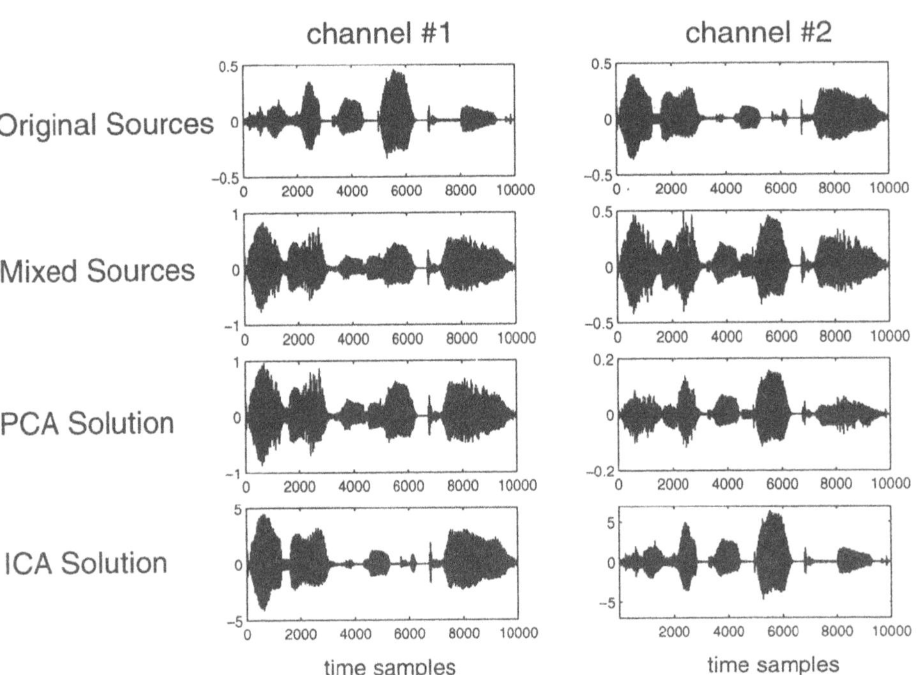

Figure 2.3. Top: the original sources, second row: the mixtures, third row: the recovered sources using PCA, bottom: the recovered sources using ICA.

matrix for the PCA solution is

$$\begin{bmatrix} 1.3706 & 2.2301 \\ 0.3483 & -0.1628 \end{bmatrix}.$$

The off-diagonal coefficients indicate that the sources have not been separated. The performance matrix for the ICA solution is

$$\begin{bmatrix} 0.088 & 15.0864 \\ 12.5293 & 0.0569 \end{bmatrix}.$$

The solution indicates that the recovered sources are permuted and scaled which can be seen in figure 2.3 (bottom).

These simple examples illustrate that decorrelation-based methods cannot be used to separate independent sources. A method is necessary that can approximately take into account all higher-order correlations and make the signals truly independent.

2.4 THE INFORMATION MAXIMIZATION APPROACH TO ICA

Nadal and Parga (1994) showed that in the low-noise case, the maximum of the mutual information between the inputs $\mathbf{x}$ and outputs $\mathbf{y}$ of a neural processor implied that the output distributions were factorial. In other words, maximizing the information transfer in a nonlinear neural network minimizes the mutual information among the outputs (factorial code) when optimization is done over both the synaptic weights and the nonlinear transfer function. Roth and Baram (1996) and Bell and Sejnowski (1995) independently derived stochastic gradient learning rules for this maximization and applied them, respectively to forecasting and time series analysis, and the blind separation of sources. Bell and Sejnowski (1995) proposed a simple learning algorithm for a feedforward neural network that blindly separates linear mixtures $\mathbf{x}$ of independent sources $\mathbf{s}$ using information maximization. They show that maximizing the joint entropy $H(\mathbf{y})$ of the output of a neural processor can approximately minimize the mutual information among the output components $y_i = g_i(u_i)$ where $g_i(u_i)$ is an invertible monotonic nonlinearity and $\mathbf{u} = \mathbf{W}\mathbf{x}$.

The joint entropy at the outputs of a neural network is

$$H(y_1, \cdots, y_N) = H(y_1) + \cdots + H(y_N) - I(y_1, \cdots, y_N), \tag{2.11}$$

where $H(y_i)$ are the marginal entropies of the outputs and $I(y_1, \cdots, y_N)$ is their mutual information. Maximizing $H(y_1, \cdots, y_N)$ consists of maximizing the marginal entropies and minimizing the mutual information. The outputs $\mathbf{y}$ are amplitude-bounded random variables and therefore the marginal entropies are maximum for a uniform distribution of y_i. Maximizing the joint entropy will also decrease $I(y_1, \cdots, y_N)$ since the mutual information is always positive. For $I(y_1, \cdots, y_N) = 0$ the joint entropy is the sum of marginal entropies

$$H(y_1, \cdots, y_N) = H(y_1) + \cdots + H(y_N). \tag{2.12}$$

The maximal value for $H(y_1, \cdots, y_N)$ is achieved when the mutual information among the bounded random variables $y_1, \cdots, y_N$ is zero and their marginal distribution is uniform. As shown below, this implies that the nonlinearity $g_i(u_i)$ has the form of the cumulative density function (c.d.f.) of the true source distribution s_i. There are now two sets of parameters that determine the maximum joint entropy: the nonlinearity $y_i = g_i(u_i)$ and the synaptic efficacies $\mathbf{W}$. Bell and Sejnowski (1995) chose the nonlinearity to be a fixed logistic function. This is equivalent to assuming a prior distribution of the sources: a super-Gaussian distribution with heavy tails and a peak centered at the mean. The only remaining parameters to adapt are the synaptic weights. They can be found by maximizing the joint entropy with respect to $\mathbf{W}$. The derivative of eq.2.11 with respect to $\mathbf{W}$ relates to the KL divergence between the multivariate uniform distribution denoted as $p_1(\mathbf{y})$ and multivariate uniform estimate $p(\mathbf{y})$ in the following form

$$\frac{\partial H(\mathbf{y})}{\partial \mathbf{W}} = \frac{\partial}{\partial \mathbf{W}}(-D(p_1(\mathbf{y})\|p(\mathbf{y}))). \tag{2.13}$$

In the limit when the transfer functions $g_i(u_i)$ and $\mathbf{W}$ are optimized the joint entropy $H(\mathbf{y})$ is maximum and $p(\mathbf{y}) = p_1(\mathbf{y})$ so that $I(\mathbf{y}) = 0$. Since $g_i(u_i)$ is an invertible mapping from u_i to y_i the KL divergence in eq.2.13 is equal to the KL divergence between the estimate of the source distribution $p(\mathbf{u})$ and the sources $p(\mathbf{s})$.

$$D(p_1(\mathbf{y})\|p(\mathbf{y})) = D(p(\mathbf{s})\|p(\mathbf{u})). \tag{2.14}$$

If the mutual information between the outputs is zero $I(y_1, \cdots, y_N) = 0$, the mutual information before the nonlinearity $I(u_1, \cdots, u_N)$ must be zero as well since the nonlinearity does not introduce any dependencies. The relation between y_i and u_i and the nonlinear transfer function is (Papoulis, 1990)

$$p(y_i) = \frac{p(u_i)}{|\frac{\partial g_i(u_i)}{\partial u_i}|}. \tag{2.15}$$

For a uniform distribution of y_i, it follows that

$$p(u_i) = |\frac{\partial g_i(u_i)}{\partial u_i}|. \tag{2.16}$$

This assumes that u_i is an independent variable with a distribution that is approximately the form of the derivative of the nonlinearity. In case of the logistic function, the appropriate p.d.f. is shown in figure 2.4 (bottom). The logistic function p.d.f. has longer tails than the Gaussian p.d.f. In fact, the logistic function p.d.f. is a rough estimate for distributions of music and speech signals (see figure 2.5) because their tails are longer and heavier than the tails of the Gaussian distribution.

Bell and Sejnowski (1995) separated mixtures of several music and speech signals using infomax with a logistic activation function. Will infomax always minimize the mutual information? Bell and Sejnowski (1995) answer this question in a thought experiment where they illustrate that when there is a mismatch between the source p.d.f. and the slope of the nonlinearity a maximal joint entropy value can be achieved with $I(\mathbf{y}) > 0$ that is higher than the joint entropy with $I(\mathbf{y}) = 0$ (due to lower

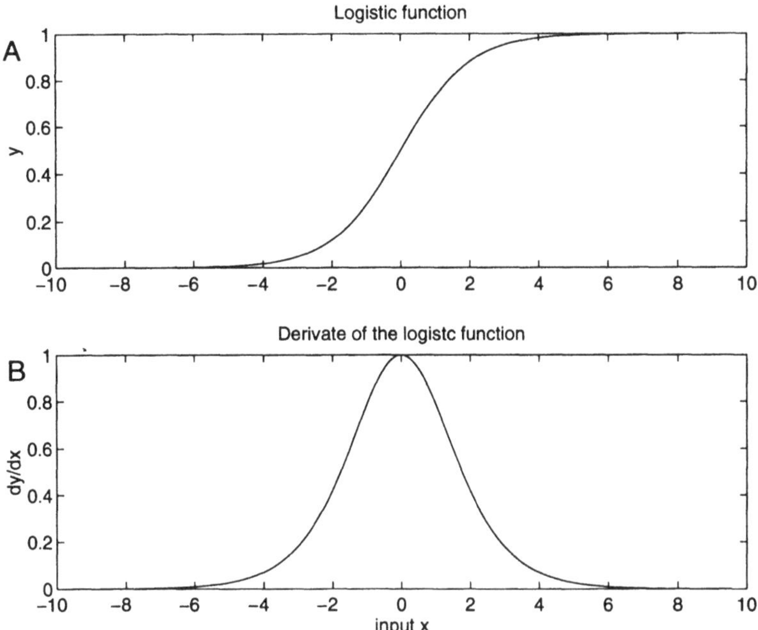

Figure 2.4. Top: logistic function and bottom: the derivate of the logistic function.

marginal entropies). In those cases, infomax will not minimize the mutual information. This is exactly the case when the mismatch between the nonlinearity and cumulative density function (c.d.f.) of the true source distribution does not satisfy the robustness criteria. An intuitive explanation is presented in section 2.8.2 for the convergence to an ICA solution although the nonlinearity does not accurately relate to the source density.

2.5 DERIVATION OF THE INFOMAX LEARNING RULE FOR ICA

In this section the derivation of the infomax learning rule by Bell and Sejnowski (1995) is summarized.

The derivation is based on a simple neural network architecture that can realize the mapping from $\mathbf{x}$ to $\mathbf{y} = g(\mathbf{u})$ is a single-layer feedforward neural network with a nonlinear output activation function. The nonlinearity $g_i(u)$ is essential for minimizing the mutual information to perform ICA. Another interpretation for the use of the nonlinearity is that it provides a combination of higher-order statistics through its Taylor series expansion that is essential to minimize higher-order correlations. The learning rule can be derived by maximizing the output entropy $H(\mathbf{y})$ of a neural processor, as proposed by Bell and Sejnowski (1995).

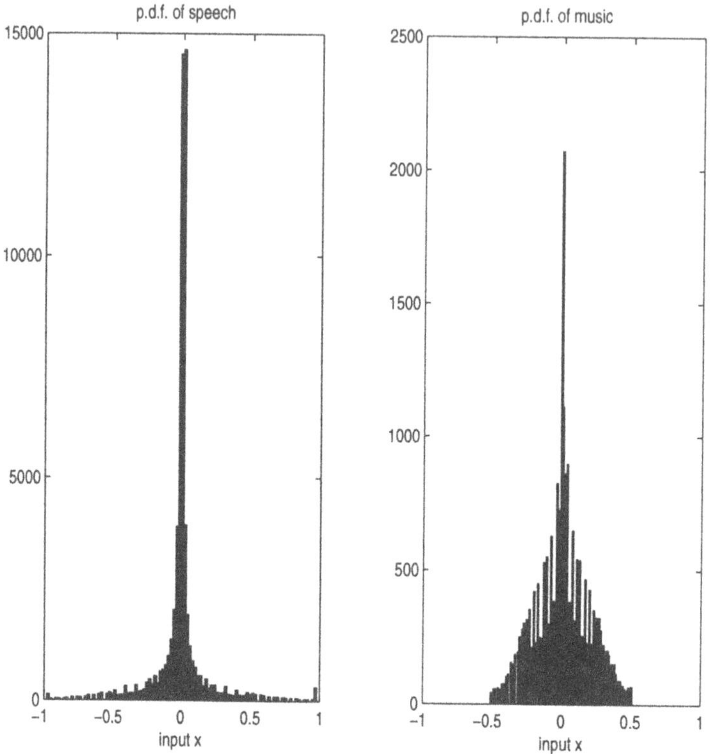

Figure 2.5. Left: p.d.f of a speech signal. Right: p.d.f. of a music signal.

The joint entropy at the outputs of a neural network is

$$H(y_1, \cdots, y_N) = H(y_1) + \cdots + H(y_N) - I(y_1, \cdots, y_N), \tag{2.17}$$

where $H(y_i)$ are the marginal entropies of the outputs and $I(y_1, \cdots, y_N)$ is their mutual information. Eq.2.17 in vector notation is

$$H(\mathbf{y}) = H(y_1) + \cdots + H(y_N) - I(\mathbf{y}). \tag{2.18}$$

Each marginal entropy can be written as

$$H(y_i) = -E\{\log p(y_i)\}. \tag{2.19}$$

The nonlinear mapping between the output density $p(y_i)$ and source estimate density $p(u_i)$ can be described by the absolute value of the derivative with respect to u_i (Papoulis, 1990)

$$p(y_i) = \frac{p(u_i)}{|\frac{\partial y_i}{\partial u_i}|}, \tag{2.20}$$

which can be substituted in eq.2.19 giving

$$H(y_i) = -E\{\log \frac{p(u_i)}{|\frac{\partial y_i}{\partial u_i}|}\}. \tag{2.21}$$

Rewriting eq.2.18 gives

$$H(\mathbf{y}) = -E\{\log \frac{p(u_1)}{|\frac{\partial y_1}{\partial u_1}|}\} + \cdots - E\{\log \frac{p(u_N)}{|\frac{\partial y_N}{\partial u_N}|}\} - I(\mathbf{y}) \tag{2.22}$$

$$H(\mathbf{y}) = -\sum_{i=1}^{N} E\{\log \frac{p(u_i)}{|\frac{\partial y_i}{\partial u_i}|}\} - I(\mathbf{y}). \tag{2.23}$$

Taking the derivative of the joint entropy is now

$$\frac{\partial H(\mathbf{y})}{\partial \mathbf{W}} = \frac{\partial}{\partial \mathbf{W}}(-I(\mathbf{y})) - \frac{\partial}{\partial \mathbf{W}} \sum_{i=1}^{N} E\{\log \frac{p(u_i)}{|\frac{\partial y_i}{\partial u_i}|}\}. \tag{2.24}$$

This equation makes a clear relation between maximizing the joint entropy of the output of the neural processor and minimizing the mutual information between the components at the outputs. A direct minimization of the mutual information is achieved when $p(u_i) = |\frac{\partial y_i}{\partial u_i}|$ is satisfied, i.e. the density of the estimated source u_i is the derivative of the nonlinear activation function y_i. In other words, the mutual information will be minimized when the nonlinearity $y_i = g_i(u_i)$ is the cumulative density function of the source estimates u_i. In case of a mismatch between the estimated source density $p(u_i)$ and the derivative of the nonlinear activation function $\frac{\partial y_i}{\partial u_i}$ the maximum of $H(\mathbf{y})$ may be achieved without $\mathbf{I}(\mathbf{y})$ being zero. In this case the error term $\frac{\partial}{\partial \mathbf{W}} \sum_{i=1}^{N} E\{\log \frac{p(u_i)}{|\frac{\partial y_i}{\partial u_i}|}\}$ exists and may interfere during the minimization process of $I(\mathbf{y})$. Therefore, eq.2.24 suggests that the minimization process of $I(\mathbf{y})$ depends on the source density estimate modeled by the nonlinearity $g_i(u_i)$. To what extend the density estimate must approximate the true source density is still an open question. Simulation results in section 2.8.2 suggest conditions under which the infomax algorithm will separate independent sources.

Assume that the error term in eq.2.24 is negligible due to the assumption that the nonlinearity $g_i(u_i)$ is flexible and able to sufficiently approximate the source density. In this case the error term vanishes and the maximum of the joint entropy $H(\mathbf{y})$ can be found by deriving $H(\mathbf{y})$ with respect to $\mathbf{W}$, i.e. computing the gradient of $H(\mathbf{y})$.

$$\frac{\partial H(\mathbf{y})}{\partial \mathbf{W}} = \frac{\partial}{\partial \mathbf{W}} \left(-E\{\log |\mathbf{J}|\}\right), \tag{2.25}$$

where the nonlinear mapping between the output density $p(\mathbf{y})$ and input density $p(\mathbf{x})$ can be described by the Jacobian (Papoulis, 1990)

$$p(\mathbf{y}) = \frac{p(\mathbf{x})}{|\mathbf{J}(\mathbf{x})|}. \tag{2.26}$$

This transformation can be seen as a volume conserving transformation (Deco and Brauer, 1995). The term $-E\{\log p(\mathbf{x})\}$ in eq.2.25 does not depend on the parameter $\mathbf{W}$. Now consider a training set of the data $\mathbf{x}$ so that the stochastic learning rule can now be approximated without the expectation term

$$\frac{\partial H(\mathbf{y})}{\partial \mathbf{W}} = \frac{\partial}{\partial \mathbf{W}} \log |\mathbf{J}|. \tag{2.27}$$

The term $|\mathbf{J}|$ is the absolute value of the Jacobian of the transformation from $\mathbf{x}$ to $\mathbf{y}$. It is the determinant of the matrix of partial derivatives

$$\mathbf{J}(\mathbf{x}) = \det \begin{bmatrix} \frac{\partial y_1}{\partial x_1} & \cdots & \frac{\partial y_1}{\partial x_n} \\ \vdots & & \vdots \\ \frac{\partial y_n}{\partial x_1} & \cdots & \frac{\partial y_n}{\partial x_n} \end{bmatrix}. \tag{2.28}$$

Considering the elements of the Jacobian, each partial derivative has the following form

$$\frac{\partial y_i}{\partial x_j} = w_{ij} \frac{\partial y_i}{\partial u_j}. \tag{2.29}$$

Since there are no connections between the outputs of the neuron the partial derivative $\partial y_i / \partial u_j$ is non-zero for $i = j$ only. Therefore the Jacobian can be rewritten as

$$\mathbf{J}(\mathbf{x}) = \det(\mathbf{W}) \prod_{i=1}^{N} |\frac{\partial y_i}{\partial u_i}|. \tag{2.30}$$

Substituting eq.2.30 in eq.2.27 it follows

$$\begin{aligned} \frac{\partial H(\mathbf{y})}{\partial \mathbf{W}} &= \frac{\partial}{\partial \mathbf{W}} \log \left(|\det(\mathbf{W})| \prod_{i=1}^{N} |\frac{\partial y_i}{\partial u_i}| \right) \\ &= \frac{\partial}{\partial \mathbf{W}} \log|\det(\mathbf{W})| + \frac{\partial}{\partial \mathbf{W}} \sum_{i=1}^{N} \log |\frac{\partial y_i}{\partial u_i}| \\ &= \frac{\partial}{\partial \mathbf{W}} \log|\det(\mathbf{W})| + \sum_{i=1}^{N} \frac{\partial}{\partial \mathbf{W}} \log |\frac{\partial y_i}{\partial u_i}|. \end{aligned} \tag{2.31}$$

The first term in eq.2.31 is

$$\frac{\partial}{\partial \mathbf{W}} \log|\det(\mathbf{W})| = \frac{(\mathrm{adj}\mathbf{W})^T}{\det \mathbf{W}} = (\mathbf{W}^T)^{-1}, \tag{2.32}$$

because the determinant of $\mathbf{W}$ can be expressed as

$$\det(\mathbf{W}) = \sum_{j=1}^{N} w_{ij} \mathrm{cof}(w_{ij}). \tag{2.33}$$

The second term in eq.2.31can be further computed as

$$\frac{\partial}{\partial w_{ij}} \sum_{i=1}^{N} \log |\frac{\partial y_i}{\partial u_i}| = \frac{1}{\frac{\partial y_i}{\partial u_i}} \frac{\partial^2 y_i}{\partial u_i^2} x_j. \tag{2.34}$$

Define the derivative of the nonlinearity y_i with respect to u_i as an approximation of the source density $p(u_i)$.

$$p(u_i) = \frac{\partial y_i}{\partial u_i}. \tag{2.35}$$

From eq.2.35 and eq.2.34 it follows that

$$\frac{\partial}{\partial \mathbf{W}} \sum_{i=1}^{N} \log |\frac{\partial y_i}{\partial u_i}| = \frac{\frac{\partial p(\mathbf{u})}{\partial \mathbf{u}}}{p(\mathbf{u})} \mathbf{x}^T. \tag{2.36}$$

Now the first and second term in eq.2.31 are computed and the learning infomax rule is (Bell and Sejnowski, 1995)

$$\frac{\partial H(\mathbf{y})}{\partial \mathbf{W}} = (\mathbf{W}^T)^{-1} + \left(\frac{\frac{\partial p(\mathbf{u})}{\partial \mathbf{u}}}{p(\mathbf{u})} \right) \mathbf{x}^T. \tag{2.37}$$

This learning rule is a result of the gradient of the entropy function and involves a computationally intensive matrix inversion.

A much more efficient way to maximize the joint entropy is to follow the 'natural' gradient. The natural gradient rescales the entropy gradient by post-multiplying the entropy gradient by $\mathbf{W}^T\mathbf{W}$ giving

$$\Delta \mathbf{W} \propto \frac{\partial H(\mathbf{y})}{\partial \mathbf{W}} \mathbf{W}^T \mathbf{W} = \left[\mathbf{I} + \left(\frac{\frac{\partial p(\mathbf{u})}{\partial \mathbf{u}}}{p(\mathbf{u})} \right) \mathbf{u}^T \right] \mathbf{W}, \tag{2.38}$$

as proposed by Amari et al. (1996), or equivalently the relative gradient by Cardoso and Laheld (1996). **I** denotes the identity matrix. An intuition about the natural gradient is in subsection 2.8.1. Furthermore define the nonlinearity

$$\varphi(\mathbf{u}) = -\frac{\frac{\partial p(\mathbf{u})}{\partial \mathbf{u}}}{p(\mathbf{u})}, \tag{2.39}$$

which is also called the score function and the equation in eq.2.38 reads

$$\Delta \mathbf{W} \propto \left[\mathbf{I} - \varphi(\mathbf{u}) \mathbf{u}^T \right] \mathbf{W}. \tag{2.40}$$

The form of $\varphi(\mathbf{u})$ plays an important role in separating sub- and super Gaussian sources because it is a function of the nonlinearity y_i and therefore a function of the source estimate. For super-Gaussian sources, Bell and Sejnowski (1995) presented a table with activation functions and their resulting nonlinearity $\varphi(\mathbf{u})$ for the learning rule.

2.6 A SIMPLE BUT GENERAL ICA LEARNING RULE

An alternative way to derive the general learning rule is given by the maximum likelihood formulation (MLE). The MLE approach to blind source separation was first proposed by Gaeta and Lacoume (1990), Pham and Garrat (1997) and was pursued more recently by Pearlmutter and Parra (1996) and Cardoso (1997). The probability density function of the observations $\mathbf{x}$ can be expressed as (Amari and Cardoso, 1997)

$$p(\mathbf{x}) = |\det(\mathbf{W})| p(\mathbf{u}), \tag{2.41}$$

where $p(\mathbf{u}) = \prod_{i=1}^{N} p_i(u_i)$ is the hypothesized distribution of $p(\mathbf{s})$. The log-likelihood of eq.2.41 is

$$L(\mathbf{u}, \mathbf{W}) = \log|\det(\mathbf{W})| + \sum_{i=1}^{N} \log p_i(u_i). \tag{2.42}$$

Maximizing the log-likelihood with respect to $\mathbf{W}$ gives a learning algorithm for $\mathbf{W}$ (Bell and Sejnowski, 1995)

$$\Delta\mathbf{W} \propto \left[(\mathbf{W}^T)^{-1} - \varphi(\mathbf{u})\mathbf{x}^T\right], \tag{2.43}$$

where

$$\varphi(\mathbf{u}) = -\frac{\frac{\partial p(\mathbf{u})}{\partial \mathbf{u}}}{p(\mathbf{u})} = \left[-\frac{\frac{\partial p(u_1)}{\partial u_1}}{p(u_1)}, \dots, -\frac{\frac{\partial p(u_N)}{\partial u_N}}{p(u_N)}\right]^T. \tag{2.44}$$

An efficient way to maximize the log-likelihood is to follow the 'natural' gradient (Amari, 1998)

$$\Delta\mathbf{W} \propto \frac{\partial L(\mathbf{u}, \mathbf{W})}{\partial \mathbf{W}} \mathbf{W}^T\mathbf{W} = \left[\mathbf{I} - \varphi(\mathbf{u})\mathbf{u}^T\right]\mathbf{W}, \tag{2.45}$$

as proposed by Amari et al. (1996) or relative gradient, Cardoso and Laheld (1996). Here $\mathbf{W}^T\mathbf{W}$ rescales the gradient, simplifies the learning rule in eq.2.43 and speeds convergence considerably. It has been shown that the general learning algorithm in eq.2.45 can be derived from several theoretical viewpoints such as MLE (Pearlmutter and Parra, 1996), infomax (Bell and Sejnowski, 1995) and negentropy maximization (Girolami and Fyfe, 1997c). Lee et al. (1998a) review these techniques and show their relation to each other.

The parametric density estimate $p_i(u_i)$ plays an essential role in the success of the learning rule in eq.2.45. Local convergence is assured if $p_i(u_i)$ is the derivative of the log-densities of the sources (Pham and Garrat, 1997). If $g_i(u)$ is chosen to be a logistic function ($g_i(u_i) = \tanh(u_i)$) so that $\varphi(\mathbf{u}) = 2\tanh(\mathbf{u})$ the learning rule reduces to that in Bell and Sejnowski (1995) with the natural gradient

$$\Delta\mathbf{W} \propto \left[\mathbf{I} - 2\tanh(\mathbf{u})\mathbf{u}^T\right]\mathbf{W}. \tag{2.46}$$

Theoretical considerations as well as empirical observations [1] have shown that this algorithm is limited to separating sources with super-Gaussian distributions. The sigmoid function used in Bell and Sejnowski (1995) provides a priori knowledge about the source distribution, i.e. the super-Gaussian shape of the sources. However, they also discuss a 'flexible' sigmoid function (a sigmoid function with parameters p, r so that $g(u_i) = \int g(u_i)^p (1 - g(u_i))^r$) can be used to match the source distribution. The idea of modeling a parametric nonlinearity has been further investigated and generalized by Pearlmutter and Parra (1996) in their contextual ICA (cICA) algorithm. They model the p.d.f. in a parametric form by taking into account the temporal information and by choosing $p_i(u_i)$ as a weighted sum of several logistic density functions with variable means and scales. Moulines et al. (1997) and Xu et al. (1997) model the underlying p.d.f. with mixtures of Gaussians and show that they can separate sub and super-Gaussian sources. These parametric modeling approaches are in general computationally expensive. In addition, our empirical results on EEG and event related potentials (ERP) using contextual ICA indicate that cICA can fail to find independent components. One possible source of error may be due to the limited number of recorded time points (e.g. 600 data points for ERPs) from which a reliable density estimate is difficult.

2.6.1 Deriving the extended infomax learning rule to separate sub- and super-Gaussian sources

The purpose of the extended infomax algorithm is to provide a simple learning rule with a fixed nonlinearity that can separate sources with a variety of distributions. One way of generalizing the learning rule to sources with either sub- or super-Gaussian distributions is to approximate the estimated p.d.f. with an Edgeworth expansion or Gram-Charlier expansion (Stuart and Ord, 1987) as proposed by Girolami and Fyfe (1997c). In Girolami (1997b) a parametric density estimate was used to derive the same learning rule without making any approximations as shown below.

A symmetric strictly sub-Gaussian density can be modeled using a symmetrical form of the Pearson mixture model (Pearson, 1894) as follows (Girolami, 1998, 1997b).

$$p(u) = \frac{1}{2}\left(N(\mu, \sigma^2) + N(-\mu, \sigma^2)\right), \tag{2.47}$$

where $N(\mu, \sigma^2)$ is the normal density with mean μ and variance σ^2. Figure 2.6 shows the form of the density $p(u)$ for $\sigma^2 = 1$ with varying $\mu = [0 \cdots 2]$. For $\mu = 0$ $p(u)$ is a Gaussian model whereas for e.g. $\mu_i = 1.5$ the $p(u)$ is clearly bimodal. The kurtosis k_4 (normalized 4th-order cumulant) of $p(u)$ is

$$\kappa = \frac{c_4}{c_2^2} = \frac{-2\mu^4}{(\mu^2 + \sigma^2)^2}, \tag{2.48}$$

where c_i is the ith-order cumulant .(Girolami, 1997b) Depending on the values of μ and σ^2 the kurtosis lies between -2 and 0. So eq.2.47 defines a strictly sub-Gaussian

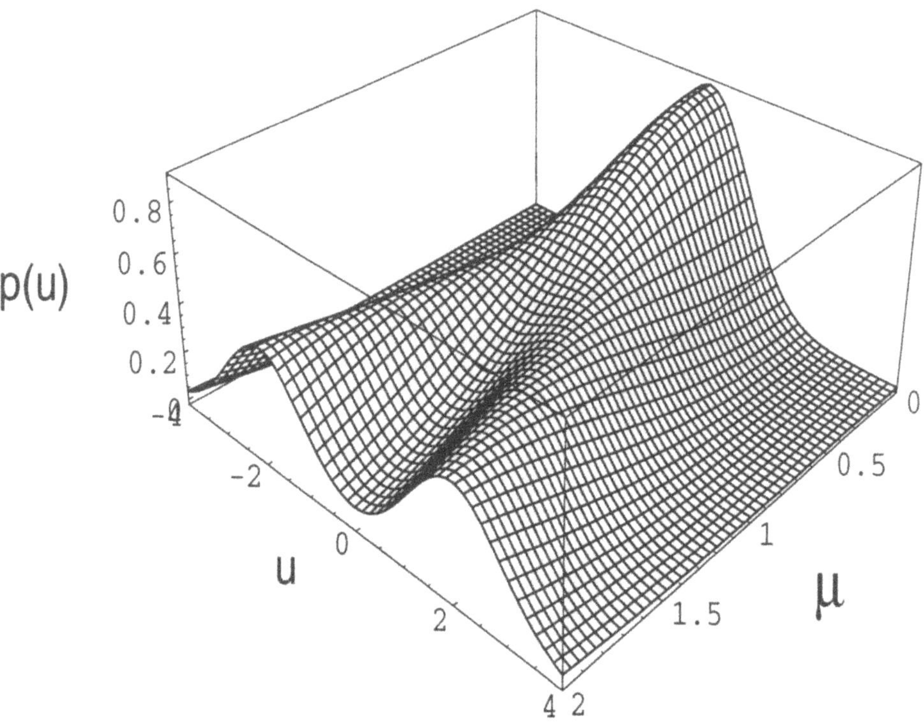

Figure 2.6. Estimated sub-Gaussian density models for the extended infomax learning rule with $\sigma^2 = 1$ and $\mu_i = \{0 \cdots 2\}$. For $\mu_i = 1.5$ the density becomes clearly bimodal.

symmetric density when $\mu > 0$. Defining $a = \frac{\mu}{\sigma^2}$ and applying eq.2.47 the term $\varphi(u)$ is now

$$\varphi(u) = -\frac{\frac{\partial p(u)}{\partial u}}{p(u)} = \frac{u}{\sigma^2} - a\left(\frac{\exp(au) - \exp(-au)}{\exp(au) + \exp(-au)}\right). \tag{2.49}$$

Using the definition of the hyperbolic tangent

$$\varphi(u) = \frac{u}{\sigma^2} - \frac{\mu}{\sigma^2}\tanh\left(\frac{\mu}{\sigma^2}u\right). \tag{2.50}$$

Setting $\mu = 1$ and $\sigma^2 = 1$ eq.2.50 reduces to

$$\varphi(u) = u - \tanh(u). \tag{2.51}$$

The learning rule for strictly sub-Gaussian sources is now (eq.2.45 and eq.2.51)

$$\Delta\mathbf{W} \propto \left[\mathbf{I} + \tanh(\mathbf{u})\mathbf{u}^T - \mathbf{u}\mathbf{u}^T\right]\mathbf{W}. \tag{2.52}$$

In the case of unimodal super-Gaussian sources the following density model is adopted

$$p(u) \propto p_G(u)\mathrm{sech}^2(u), \tag{2.53}$$

where $p_G(u) = N(0,1)$ is a zero-mean Gaussian density with unit variance. Figure 2.7 shows the density model for $p(u)$. The nonlinearity $\varphi(u)$ is now

$$\varphi(u) = -\frac{\frac{\partial p(u)}{\partial u}}{p(u)} = u + \tanh(u). \tag{2.54}$$

The learning rule for super-Gaussian sources is (eq.2.45 and eq.2.54)

$$\Delta\mathbf{W} \propto \left[\mathbf{I} - \tanh(\mathbf{u})\mathbf{u}^T - \mathbf{u}\mathbf{u}^T\right]\mathbf{W}. \tag{2.55}$$

The difference between the super-Gaussian learning rule in eq.2.55 and the sub-Gaussian learning rule eq.2.52 is the sign before the tanh-function:

$$\Delta\mathbf{W} \propto \begin{cases} \left[\mathbf{I} - \tanh(\mathbf{u})\mathbf{u}^T - \mathbf{u}\mathbf{u}^T\right]\mathbf{W} & : \ \mathrm{super-Gaussian} \\ \left[\mathbf{I} + \tanh(\mathbf{u})\mathbf{u}^T - \mathbf{u}\mathbf{u}^T\right]\mathbf{W} & : \ \mathrm{sub-Gaussian} \end{cases} \tag{2.56}$$

The learning rules differ in the sign before the tanh-function and can be determined using a switching criterion. Girolami (1997b) employs the sign of kurtosis of the analysis as a switching criterion. However, as there is no general definition for sub- and super-Gaussian sources a valid choice for a switching criterion is based on stability criteria which is presented in the next subsection.

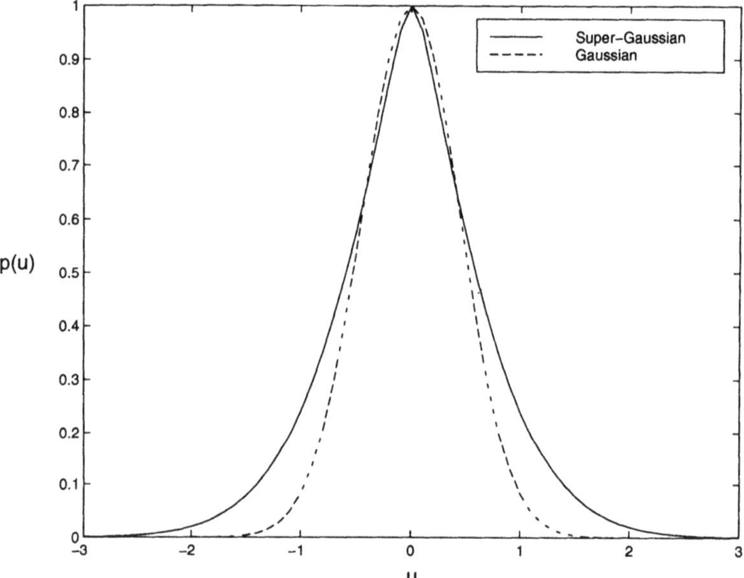

Figure 2.7. Density model for the super-Gaussian distribution. The super-Gaussian model has a heavier tail than the normal density.

2.6.2 Switching between nonlinearities

The switching between the sub- and super-Gaussian learning rule is

$$\Delta \mathbf{W} \propto \left[\mathbf{I} - \mathbf{K}\tanh(\mathbf{u})\mathbf{u}^T - \mathbf{u}\mathbf{u}^T\right]\mathbf{W} \begin{cases} k_i = 1 & : \ \text{super} - \text{Gaussian} \\ k_i = -1 & : \ \text{sub} - \text{Gaussian} \end{cases} \tag{2.57}$$

where k_i are elements of the N-dimensional diagonal matrix $\mathbf{K}$. The switching moments k_i can be derived from the generic stability analysis of separating solutions as employed by Cardoso and Laheld (1996) [2], Pham and Garrat (1997) and Amari et al. (1997a). In the stability analysis the mean field is approximated by a first-order perturbation in the parameters of the separating matrix. The linear approximation near the stationary point is the gradient of the mean field at the stationary point. The real part of the eigenvalues of the derivative of the mean field must be negative so that the parameters are on average pulled back to the stationary point

A sufficient condition guaranteeing asymptotic stability can be derived (Cardoso, 1998b, 1998c) so that

$$\kappa_i > 0 \qquad 1 \leq i \leq N \tag{2.58}$$

where κ_i is

$$\kappa_i = E\{\varphi_i'(u_i)\}E\{u_i^2\} - E\{\varphi_i(u_i)u_i\}, \tag{2.59}$$

and

$$\varphi_i(u_i) = u_i + k_i \tanh(u_i). \tag{2.60}$$

Substituting eq.2.60 in eq.2.59 gives

$$\begin{aligned} \kappa_i &= E\{k_i \text{sech}^2(u_i) + 1\}E\{u_i^2\} - E\{[k_i \tanh(u_i) + u_i]u_i\} & (2.61) \\ &= k_i \left(E\{\text{sech}^2(u_i)\}E\{u_i^2\} - E\{[\tanh(u_i)]u_i\}\right). & (2.62) \end{aligned}$$

To ensure $\kappa_i > 0$ the sign of k_i must be the same as the sign of $E\{\text{sech}^2(u_i)\}E\{u_i^2\} - E\{[\tanh(u_i)]u_i\}$. Therefore the learning rule in eq.2.57 is used where the k_i's are

$$k_i = \text{sign}\left(E\{\text{sech}^2(u_i)\}E\{u_i^2\} - E\{[\tanh(u_i)]u_i\}\right). \tag{2.63}$$

2.6.3 The hyperbolic-Cauchy density model

Another parametric density model is presented that may be used for the separation of sub- and super-Gaussian sources. Define the parametric mixture density as

$$p(u) \propto \text{sech}^2(u + b) + \text{sech}^2(u - b). \tag{2.64}$$

Figure 2.8 shows the parametric density as a function of b. For $b = 0$ the parametric density is proportional to the hyperbolic-Cauchy distribution and is therefore suited for separating super-Gaussian sources. For $b = 2$ the parametric density estimator

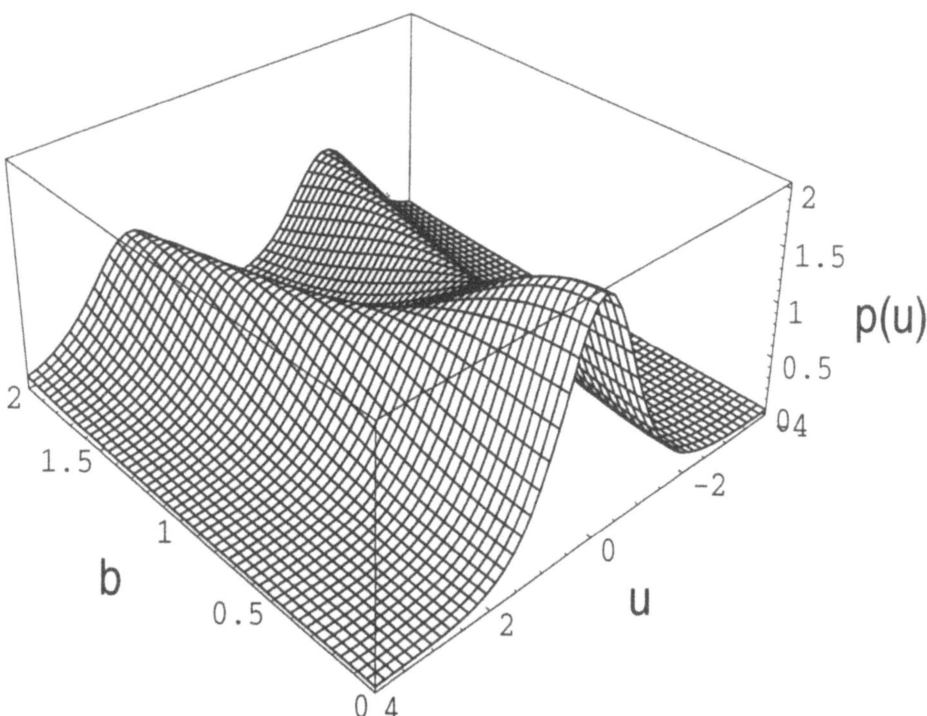

Figure 2.8. $p(u)$ as a function of b. For $b = 0$ the density estimate is suited to separate super-Gaussian sources. If for example $b = 2$ the density estimate is bimodal and therefore suited to separated sub-Gaussian sources.

has a bimodal [3] distribution with negative kurtosis and is therefore suitable for separating sub-Gaussian sources

$$\varphi(u) = -\frac{\partial}{\partial u} \log p(u) = -2\tanh(u) + 2\tanh(u+b) + 2\tanh(u-b). \tag{2.65}$$

The learning algorithm for sub- and super-Gaussian sources is now (eq.2.65 and eq.2.45)

$$\Delta \mathbf{W} \propto \left[\mathbf{I} + 2\tanh(\mathbf{u})\mathbf{u}^T - 2\tanh(\mathbf{u}+\mathbf{b})\mathbf{u}^T - 2\tanh(\mathbf{u}-\mathbf{b})\mathbf{u}^T\right] \mathbf{W}. \tag{2.66}$$

When $\mathbf{b} = \mathbf{0}$ (where $\mathbf{0}$ is a N-dim. vector with elements 0) then the learning rule reduces to

$$\Delta \mathbf{W} \propto \left[\mathbf{I} - 2\tanh(\mathbf{u})\mathbf{u}^T\right] \mathbf{W}, \tag{2.67}$$

which is exactly the learning rule in Bell and Sejnowski (1995) with the natural gradient extension. For $\mathbf{b} > \mathbf{1}$, the parametric density is bimodal (as shown in figure 2.8) and the learning rule is suitable for separating signals with sub-Gaussian distributions. Here again the sign of the general stability criteria may be used in eq.2.58 and κ_i in eq.2.59 to determine b_i so that the learning rule can switch between $b_i = 0$ and for example $b_i = 2$. Figure 2.9 shows a comparison between the range of kurtosis values of the parametric mixture density models in eq.2.47 and eq.2.64. The kurtosis value is shown as a function of the shaping parameter μ for the symmetric Pearson density model and b for the hyperbolic-Cauchy mixture density model. The kurtosis for the Pearson model is strictly negative except for $\mu = 0$ when the kurtosis is zero. Because the kurtosis for the hyperbolic-Cauchy model ranges from positive to negative, it may be used to separate signals with both sub- and super-Gaussian densities.

2.7 SIMULATIONS

Extensive simulations and experiments were performed on recorded data to verify the performance of the extended infomax algorithm eq.2.56. The results demonstrate that the algorithm is able to separate a large number of sources with a wide variety of sub- and super-Gaussian distributions. The performance of the extended infomax learning rule in eq.2.45 is compared to the original infomax learning rule eq.2.46.

2.7.1 10 Mixed Sound Sources

Ten mixed sound sources were obtained which were separated by contextual ICA as described in Pearlmutter and Parra (1996). No prewhitening is required since the transformation $\mathbf{W}$ is not restricted to a rotation in contrast to nonlinear PCA (Karhunen et al., 1997c). All 55000 data points were passed 20 times through the learning rule using a block size (batch) of 300. This corresponds to 3666 iterations (weight updates). The learning rate was fixed at 0.0005. Figure 2.10 shows the error

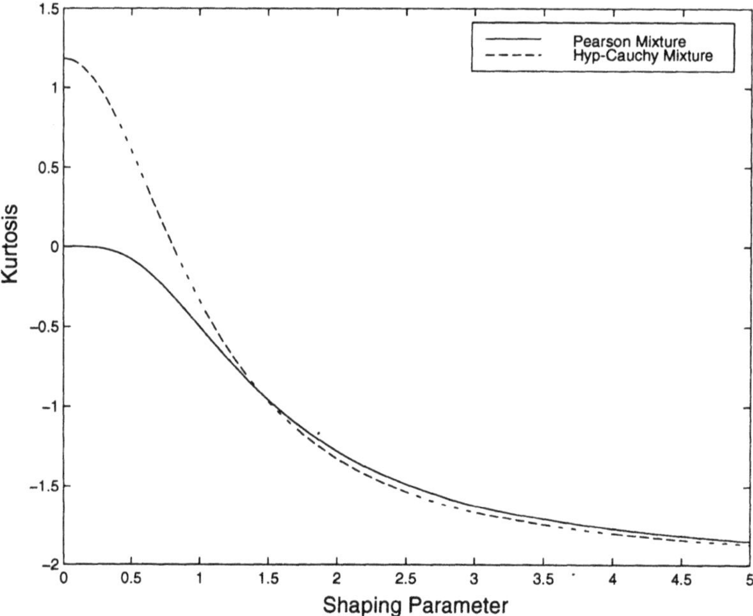

Figure 2.9. The kurtosis value is shown as a function of the shaping parameter μ and b (μ for the Pearson density model and b for the hyperbolic-Cauchy density model). Both models approach $k_4 = -2$ as the shaping parameter increases. The kurtosis for the Pearson model is strictly negative except for $\mu = 0$. The kurtosis for the hyperbolic-Cauchy model ranges from positive to negative so that we may use this single parametric model to separate signals with sub- and super-Gaussian densities.

Table 2.1. Simulation results with 10 sources

-0.09	-0.38	0.14	-0.10	-0.06	0.93	-0.36	-0.54	0.17	14.8
-11.2	-0.01	0.14	0.05	-0.08	0.02	0.07	0.21	-0.12	-0.68
0.15	0.08	-0.08	-0.02	10.2	-0.02	0.15	0.05	0.07	0.17
0.39	0.61	-0.70	-0.07	0.14	0.32	-0.08	0.85	7.7	-0.16
0.04	0.76	14.9	0.03	0.03	-0.17	0.18	-0.31	-0.19	0.04
0.11	12.9	-0.54	-0.23	-0.43	-0.21	-0.12	0.05	0.07	0.18
0.45	0.16	-0.02	6.5	0.24	0.98	-0.39	-0.97	0.06	-0.08
0.31	0.14	0.23	0.03	-0.14	-17.3	-0.39	-0.25	0.19	0.39
-0.54	-0.81	0.62	0.84	-0.18	0.47	-0.04	10.5	-0.92	0.12
-0.08	-0.26	0.15	-0.10	0.49	0.01	-10.3	0.59	0.33	-0.94

The performance matrix **P** for 10 mixed sound sources after one pass through the data. **P** is already close to the identity matrix after rescaling and reordering.

measure during learning. Both learning rules converged. The small variations of the extended infomax algorithm (upper curve) were due to the adaptation process of **K**. The matrix **K** was initialized to the identity matrix and during the learning process the elements of **K** converge to -1 or 1 to extract sub- or super-Gaussian sources respectively. In this simulation example, sources 7,8 and 9 are close to Gaussian and slight variations of their density estimation change the sign. Annealing of the learning rate reduced the variation (Lee and Sejnowski, 1997). All the music signals had super-Gaussians distribution and therefore were separable by the original infomax algorithm. The sources are already well separated after one pass through the data (about 10 sec on a Sparc 10 workstation using MATLAB) as shown in table 2.1:

For all experiments and simulations, a momentum term helped to accelerate the convergence of the algorithm

$$\Delta \mathbf{W}(n+1) = (1-\alpha)\Delta \mathbf{W}(n) + \alpha \mathbf{W}(n), \tag{2.68}$$

where α takes into account the history of **W** and α can be increased with increasing number of weight updates (as $n \to \infty$, $\alpha \to 1$).

The performance during the learning process was monitored by the error measure that was proposed by Amari et al. (1996)

$$E = \sum_{i=1}^{N} \left(\sum_{j=1}^{N} \frac{|p_{ij}|}{\max_k |p_{ik}|} - 1 \right) + \sum_{j=1}^{N} \left(\sum_{i=1}^{N} \frac{|p_{ij}|}{\max_k |p_{kj}|} - 1 \right), \tag{2.69}$$

where p_{ij} are elements of the performance matrix $\mathbf{P} = \mathbf{WA}$. **P** is close to a permutation of the scaled identity matrix when the sources are separated. Figure 2.10 shows the error measure during the learning process.

The observation in this simulation is that compared to contextual ICA which converged after several hundred passes (Pearlmutter 1997; personal communication) the original infomax algorithm shows faster convergence.

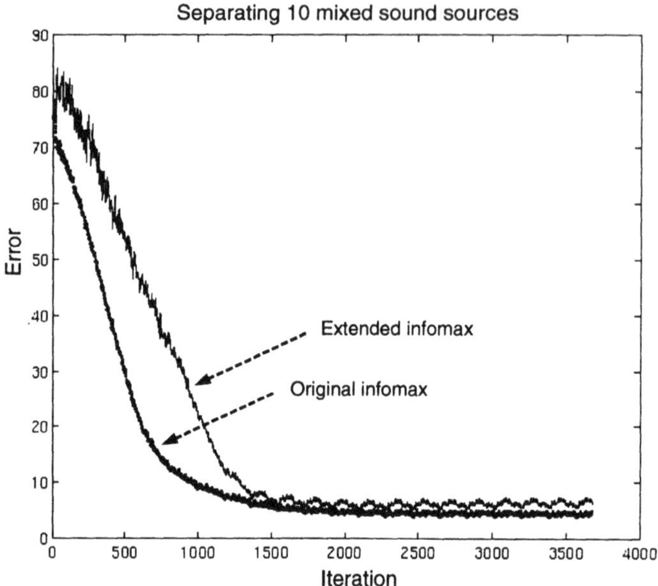

Figure 2.10. Error measure E in eq.2.69 for the separation of 10 sound sources. The upper curve is the performance for extended infomax and the lower curve shows the performance for the original infomax.

To compare the speed of the extended infomax algorithm with another closely related ones, the 10 mixed sound sources were separated using the extended exploratory projection pursuit network with inhibitory lateral connections Girolami and Fyfe (1997b). The single feedforward neural network converged several times faster than this architecture using the same learning rate and a block size of 1. Larger block sizes can be used in the feedforward network but not the feedback networks, which increases the convergence speed considerably due to a more reliable estimate of the switching matrix **K**.

2.7.2 20 Mixed Sound Sources

In this simulation experiment 20 sources of the following were separated: 10 sound tracks obtained from Pearlmutter, 6 speech & sound signals used in Bell and Sejnowski (1995), 3 uniformly distributed sub-Gaussian noise signals and one noise source with a Gaussian distribution. The kurtosis of the 20 sources are shown in table 2.2. The densities of the mixtures were close to the Gaussian distributions. The following parameters were used: learning rate fixed at 0.0005, block size of 100 data points, 150 passes through the data (41250 iterations).

Figure 2.11 shows the performance of the matrix **P** after the rows were manually reordered and normalized to unity. **P** is close to the identity matrix and its off diagonal elements indicate the amount of error. In this simulation k_4 is employed as

Table 2.2. Simulation results with 20 sources

Source number	Source type	Original kurtosis	Recovered kurtosis, infomax	Recovered kurtosis, ext. infomax	SNR for ext. infomax
1	Music 1	2.4733	2.4754	2.4759	43.4
2	Music 2	1.5135	1.5129	1.5052	55.2
3	Music 3	2.4176	2.4206	2.4044	44.1
4	Music 4	1.076	1.0720	1.0840	31.7
5	Music 5	1.0317	1.0347	1.0488	43.6
6	Music 6	1.8626	1.8653	1.8467	48.1
7	Music 7	0.7867	0.8029	0.7871	32.7
8	Music 8	0.4639	0.2753	0.4591	29.4
9	Music 9	0.5714	0.5874	0.5733	36.4
10	Music 10	2.6358	2.6327	2.6343	46.4
11	Speech 1	6.6645	6.6652	6.6663	54.3
12	Speech 2	3.3355	3.3389	3.3324	50.5
13	Music 11	1.1082	1.1072	1.1053	48.1
14	Speech 3	7.2846	7.2828	7.2875	50.5
15	Music 12	2.8308	2.8198	2.8217	52.6
16	Speech 4	10.8838	10.8738	10.8128	57.1
17	Uni. Noise 1	-1.1959	-0.2172	-1.1955	61.4
18	Uni. Noise 2	-1.2031	-0.2080	-1.2013	67.7
19	Uni. Noise 3	-1.1966	-0.2016	-1.1955	63.6
20	Gauss. Noise	-0.0148	-0.0964	-0.0399	24.9

Kurtosis of the 20 original signal sources and the kurtosis of the recovered signals from original infomax and extended infomax. The source signals range from highly kurtotic speech signals, Gaussian noise (kurtosis is zero) to noise sources with uniform distribution (negative kurtosis). Boxes are placed around sources that failed to clearly separate. In addition, the SNR is computed for extended infomax.

a measure of the recovery of the sources. The original infomax algorithm separated most of the positive kurtotic sources. However, it failed to extract several sources including two super-Gaussian sources (music 7 & 8) with low kurtosis (0.78 and 0.46 respectively). In contrast, figure 2.12 shows that the performance matrix **P** for the extended infomax algorithm is close to the identity matrix. In a listening test, there was a clear separation of all sources from their mixtures. Note that although the sources ranged from Laplacian distributions ($p(s) \propto \exp(-|s|)$, e.g. speech), Gaussian noise, to uniformly distributed noise, they were all separated using one nonlinearity.

The simulation results suggest that the super-Gaussian and sub-Gaussian density estimates in eq.2.47 and eq.2.53 are sufficient to separate the true sources. The learning algorithms in eq.2.56 and eq.2.66 performed almost identically.

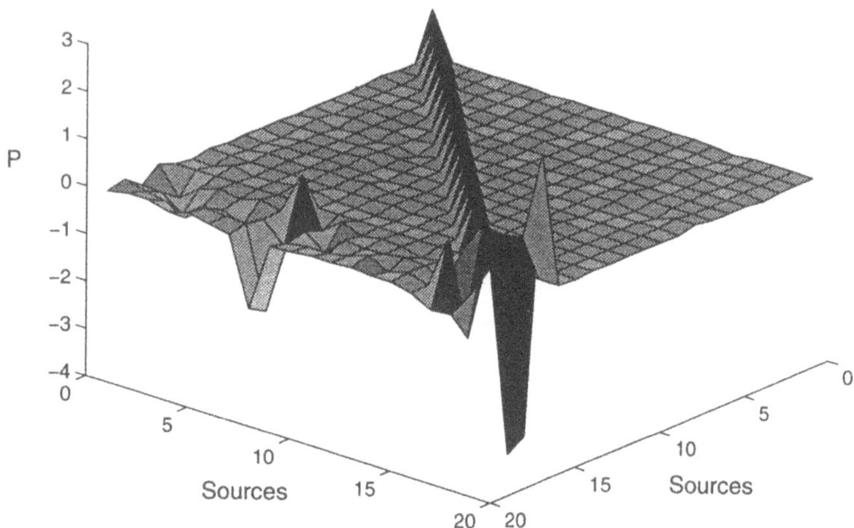

Figure 2.11. Performance matrix $\mathbf{P}$ for the separation of 20 sources using the original infomax algorithm after normalizing and reordering. Most super-Gaussian sources were recovered. However, the three sub-Gaussian sources (17,18,19), the Gaussian source (20) and two super-Gaussian sources (7, 8) remain mixed and alias in other sources. In total, 14 sources were extracted and 6 channels remained mixed (see Table 2).

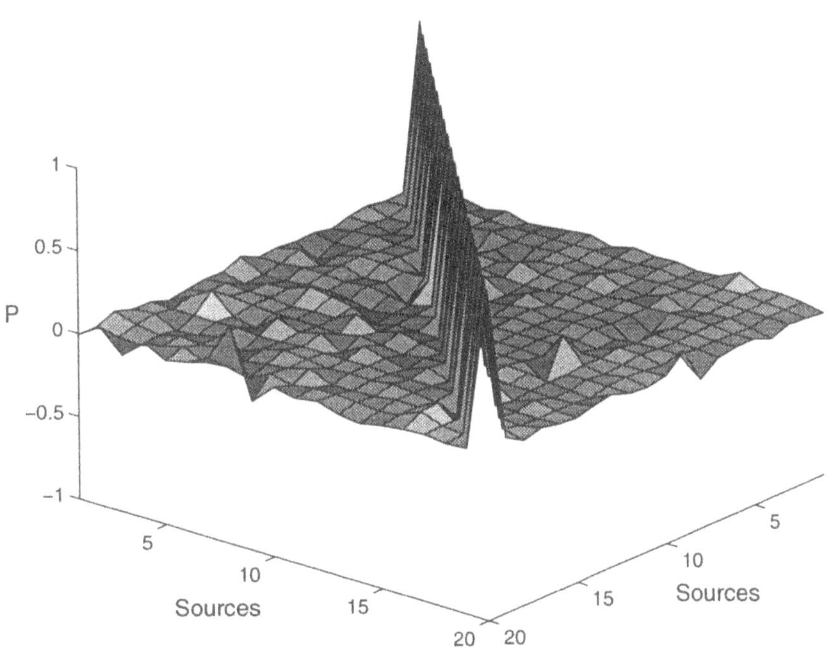

Figure 2.12. Performance matrix $\mathbf{P}$ for the separation of 20 sources using the extended infomax algorithm after normalizing and reordering. $\mathbf{P}$ is approximately the identity matrix which indicates nearly perfect separation.

2.8 CONVERGENCE PROPERTIES IN BLIND SOURCE SEPARATION

In this section, intuitive explanations are presented for two important convergence properties of the general learning rule in eq.2.45 when applied to source separation: (1) The natural gradient (Amari, 1998) or relative gradient (Cardoso and Laheld, 1996) property and (2) the robustness property in blind source separation. The former optimizes the convergence speed and the latter implies that a rough density estimation via the nonlinearity in the learning rule is sufficient to extract a variety of source distributions under certain conditions.

2.8.1 An intuition for the natural gradient

The natural gradient (Amari, 1998) or relative gradient (Cardoso and Laheld, 1996) is an important improvement to the blind source separation problem. The theory is presented in Amari et al. (1996); Amari (1997a, 1998) from an information geometry viewpoint. Here, an intuitive explanation is presented giving the principal thoughts leading to the derivation of the natural gradient.

The units of the learning algorithm on the left and right side of eq.2.37 do not match and hence the rate of convergence depends on the scales of the axes. The natural gradient (Amari, 1997a, 1998) or relative gradient (Cardoso and Laheld, 1996) greatly improves convergence of ICA by making the gradient invariant to the scale on the axes.

The normal entropy gradient (Euclidean gradient) assumes that the space of $\mathbf{W}$ is orthonormal, that is, each $\mathbf{w}_{ij}$ is of unit length and points in an orthogonal direction to the others. In this case the metric tensor is the identity matrix: $\mathbf{w}_{ij}.\mathbf{w}_{kl} = \delta_{(ij)(kl)}$. This space ts called the $\mathbf{I}$-space.

The space of non-orthonormal and non-singular ($\det(\mathbf{W}) \neq 0$) matrices $\mathbf{W}$ is called the $\mathbf{W}$-space. An important property of this space is that the entropy gradient with respect to $\mathbf{W}$ can move along an arbitrary curved manifold. The entropy gradient is given by the differential of the curved space, say $f(x)$. For a scalar function $f(x)$ the Taylor expansion of the gradient (differential of $f(x)$) is

$$f(x + \Delta x) - f(x) = \frac{\partial f(x)}{\partial x} \Delta x. \tag{2.70}$$

Differential geometry (Aris, 1962) can be used to extend eq.2.70 to the matrices.

$$f(\mathbf{W} + \Delta \mathbf{W}) - f(\mathbf{W}) = \langle \tilde{\nabla} f(\mathbf{W}), \Delta \mathbf{W} \rangle_{\mathbf{W}}, \tag{2.71}$$

where the gradient operator $\tilde{\nabla} f(\mathbf{W})$ is sought that is natural to $\mathbf{W}$. i.e. the metric in space $\mathbf{W}$ of the natural gradient is not depending on the point in $\mathbf{W}$. This is further described using the subscript $\mathbf{W}$. This space will not necessarily be Euclidean, i.e have orthogonal basis vectors. Define a space of matrices with non-singular orthonormal matrices $\mathbf{W}$ called the space of identity matrices ($\mathbf{I}$ space). The Euclidean space (such as a Cartesian frame for reference) is an example where the matrix differential is now

$$f(\mathbf{W} + \Delta \mathbf{W}) - f(\mathbf{W}) = \langle \nabla f(\mathbf{W}), \Delta \mathbf{W} \rangle_{\mathbf{I}}. \tag{2.72}$$

Regarding the convergence property of the gradient in each space, it is required that they are the same. The difference is the metric. Figure 2.13 shows that the convergence property for the Euclidean space is faster than the non-Euclidean space $\mathbf{W}$ for the same metric. To find the metric for the non-orthonormal space of $\mathbf{W}$ that may be used in the Euclidean space, it is necessary to rescale the Euclidean gradient so that it equals the natural gradient. The rescaling factor is found by setting the space $\mathbf{W}$ and $\mathbf{I}$ equal to each other after the space $\mathbf{W}$ has been mapped from $\mathbf{W}$ to $\mathbf{I}$ using the transformation $\mathbf{W}^{-1}$. Amari (1998) showed that the metric of the inner product $\langle\tilde{\nabla}f(\mathbf{W}),\Delta\mathbf{W}\rangle_{\mathbf{W}}$ is equivalent to the inner product of $\tilde{\nabla}f(\mathbf{W})\mathbf{W}^{-1}$ at $\mathbf{W}\mathbf{W}^{-1}$ for *any* $\mathbf{W}^{-1}$. This is due to the Riemannian structure of the space of $N \times N$ nonsingular matrices. The Riemannian structure is a differentiable manifold (hyperplane) with a positive definite metric $\mathbf{W}$. The right side of eq.2.71 can now be written as follows

$$\langle\tilde{\nabla}f(\mathbf{W}),\Delta\mathbf{W}\rangle_{\mathbf{W}} = \langle\tilde{\nabla}f(\mathbf{W})\mathbf{W}^{-1},\Delta\mathbf{W}\mathbf{W}^{-1}\rangle_{\mathbf{W}\mathbf{W}^{-1}}. \tag{2.73}$$

Now define the space of matrices in eq.2.72 to be the same as in eq.2.73 so that

$$\langle\tilde{\nabla}f(\mathbf{W})\mathbf{W}^{-1},\Delta\mathbf{W}\mathbf{W}^{-1}\rangle_{\mathbf{I}} \equiv \langle\nabla f(\mathbf{W}),\Delta\mathbf{W}\rangle_{\mathbf{I}}. \tag{2.74}$$

The inner product of matrices is equivalent to the trace function

$$\mathrm{tr}\left[\mathbf{W}^{-T}\tilde{\nabla}f(\mathbf{W})^T\Delta\mathbf{W}\mathbf{W}^{-1}\right] = \mathrm{tr}\left[\nabla f(\mathbf{W})^T\Delta\mathbf{W}\right]. \tag{2.75}$$

Since $\mathrm{tr}\,[\mathbf{AB}] = \mathrm{tr}\,[\mathbf{BA}]$ eq.2.75 is also

$$\begin{aligned} \mathrm{tr}\left[\mathbf{W}^{-1}\mathbf{W}^{-T}\tilde{\nabla}f(\mathbf{W})^T\Delta\mathbf{W}\right] &= \mathrm{tr}\left[\nabla f(\mathbf{W})^T\Delta\mathbf{W}\right] \\ \mathrm{tr}\left[\mathbf{W}^{-1}\mathbf{W}^{-T}\tilde{\nabla}f(\mathbf{W})^T\Delta\mathbf{W}\right] - \mathrm{tr}\left[\nabla f(\mathbf{W})^T\Delta\mathbf{W}\right] &= 0 \\ \mathrm{tr}\left[(\mathbf{W}^{-1}\mathbf{W}^{-T}\tilde{\nabla}f(\mathbf{W})^T - \nabla f(\mathbf{W})^T)\Delta\mathbf{W}\right] &= 0. \end{aligned} \tag{2.76}$$

Since $\Delta\mathbf{W}$ is arbitrary the remaining matrix which left multiply must equal the zero matrix. It follows therefore that

$$\begin{aligned} \mathbf{W}^{-1}\mathbf{W}^{-T}\tilde{\nabla}f(\mathbf{W})^T &= \nabla f(\mathbf{W})^T \\ \tilde{\nabla}f(\mathbf{W})^T &= \mathbf{W}^T\mathbf{W}\nabla f(\mathbf{W})^T \\ \tilde{\nabla}f(\mathbf{W}) &= \nabla f(\mathbf{W})\mathbf{W}^T\mathbf{W}. \end{aligned} \tag{2.77}$$

The natural gradient operator within the parameter space of arbitrary square matrices is equivalent to the Euclidean gradient operator post multiplied by the matrix transpose and matrix product. The convergence performance will then be the most efficient. A more detailed derivation of this intuitive explanation is presented by Yang and Amari (1997).

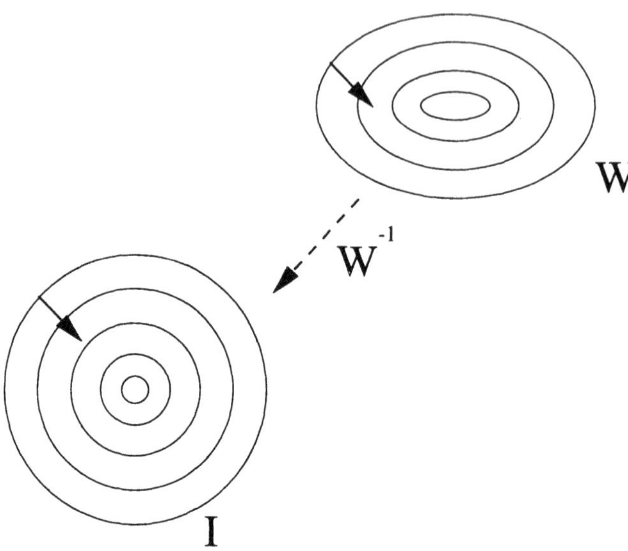

Figure 2.13. For a fixed metric, the gradient in the orthonormal space exhibits optimal convergence because it points to the center of the solution. For non orthogonal solutions a mapping into an orthonormal solution provides the most efficient performance.

2.8.2 Robustness to parameter mismatch

The insights in sections 2.4 suggest that the estimation of the true source densities should be crucial to extract the sources. Many researchers have therefore tried to find the separating matrix $\mathbf{W}$ as well as a parametric estimate of the nonlinearity (Pearlmutter and Parra, 1996; Moulines et al., 1997; Xu et al., 1997). Pearlmutter and Parra (1996) proposed a contextual ICA (cICA) algorithm that assumed a weighted sum of parametric logistic functions to model the source density. Moulines et al. (1997) and Xu et al. (1997) model the underlying p.d.f. with mixtures of Gaussians showing that they can also separate sub and super-Gaussian sources. These parametric modeling approaches in general are computationally expensive. In addition, empirical results by Lee et al. (1998b) and Makeig (personal communication) on electroencephalographic (EEG) data and data from event related potentials (ERP) using cICA indicate that it can fail to find independent components when the number of time-samples is too small to give a reliable density estimate, e.g. 600 data points for ERPs.

However, simulation results performed by many researchers show that ICA algorithms with a fixed nonlinearity converge to a separating solution although the nonlinearity implies a crude approximation of the underlying sources. Bell and Sejnowski (1995) report that the infomax algorithm can separate 10 super-Gaussian sources such as music and speech using only one logistic function that imposes a simple super-Gaussian prior. Lee et al. (1998b) report that the 10 sound sources used by Pearlmutter and Parra (1996) can be separated easily and with faster convergence than cICA using the logistic function instead of a parametric density estimator. Amari (1997b) calls this 'superefficiency' because one can extract the sources surprisingly well in simulations and real data experiments. He shows that under certain conditions a rough estimator will give a sufficient solution [4] suggesting that the covariance of the estimation error decreased on the order of $1/t^2$, t being the number of time-points. Unfortunately, decreasing the error of the covariance does not imply that the error variance of the extracted signals decreases. Cardoso (1997) suggests intuitively that a model-mismatch will still converge to a satisfactory solution because sources may be recovered up to scaling factors.

Simulation results in section 2.7 indicate that in presence of only one class of sources (either sub- or super-Gaussian sources) the algorithm using a single fixed nonlinearity converge to a separated solution. Furthermore, several simulations were performed in this section where mixtures of a wide range of symmetrical super-Gaussian source distributions were separated using different nonlinearities that deliberately showed a mismatch between the source density estimate and the randomly generated source priors. The observation is that the algorithm always converged to a separating solution while the speed of convergence and the scale of the separated sources were different. The same observations were made for sub-Gaussian sources.

It is therefore assumed that the following are sufficient conditions for the algorithm to exhibit robustness to parameter mismatch and convergence to a separating solution when the p.d.f.s of the sources $\mathbf{s}$ belong to only one class of sources (sub-Gaussian or super-Gaussian density).

1. The extended infomax learning rule is used

$$\Delta \mathbf{W} \propto \begin{cases} \left[\mathbf{I} + f(\mathbf{u})\mathbf{u}^T - \alpha \mathbf{u}\mathbf{u}^T\right]\mathbf{W} & : \text{super} - \text{Gaussian} \\ \left[\mathbf{I} - f(\mathbf{u})\mathbf{u}^T - \alpha \mathbf{u}\mathbf{u}^T\right]\mathbf{W} & : \text{sub} - \text{Gaussian} \end{cases} \tag{2.78}$$

2. $f(u_i)$ is a monotone nonlinearity and has the form of the derivative of the log-density of $p(s_i)$ where $p(s_i)$ has the form of an arbitrary symmetrical super-Gaussian distribution.

Condition (2) is similar to Amari (1997b) condition on superefficiency that holds when $f(u_i)$ is an odd function and $p(s_i)$ is an even function.

Consider the case when $f(u_i) = \frac{\partial p(u_i)/\partial u_i}{p(u_i)}$. When $\alpha = 0$ and $p(u_i)$ is the derivative of the logistic function, the learning rule in eq.2.78 reduces to the infomax learning rule as proposed by Bell and Sejnowski (1995). For $\alpha = 1$ and $p(u_i) = \text{sech}(u_i)$ the the learning rule in eq.2.78 reduces to eq.2.56. The constant α is not critical for super-Gaussian sources. However, for sub-Gaussians it is necessary that $\alpha > 0$ to satisfy the approximations for sub-Gaussian sources. The robustness formulation for eq.2.78 now requires that $f(.)$ is an odd nonlinear function of any symmetric super-Gaussian density (even functions for s_i) when only one class of sources are observed.

Several simulations were performed to verify the robustness in ICA. To this end, ten zero-mean white sources with symmetrical distributions were generated ranging from highly super-Gaussian distribution (high kurtosis) to a Gaussian distribution

Table 2.3. Robustness testing simulation with 10 sources

Source number	Original kurtosis	recovered kurtosis $f_1(\mathbf{u})$	recovered kurtosis $f_2(\mathbf{u})$	recovered kurtosis $f_3(\mathbf{u})$
1	24.478	24.4691	24.4541	24.4560
2	15.538	15.5432	15.5403	15.5327
3	13.508	13.5054	13.5019	13.4920
4	8.0094	8.0083	8.0065	8.0001
5	5.5211	5.5274	5.5258	5.5174
6	4.9338	4.9212	4.9410	4.9243
7	2.8468	2.8489	2.8482	2.8462
8	1.7084	1.7052	1.7086	1.7051
9	0.9305	0.9286	0.9277	0.9250
10	0.4691	0.4645	0.4639	0.4637

Ten zero-mean white sources with symmetrical distributions ranging from highly super-Gaussian distribution (high kurtosis) to a Gaussian distribution. Sources are recovered using $f_1(\mathbf{u}) = \text{sign}(\mathbf{u})$, $f_2(\mathbf{u}) = \tanh(\mathbf{u})$ and $f_3(\mathbf{u}) = \text{abs}(\mathbf{u}^{0.9}) \times \text{sign}(\mathbf{u})$.

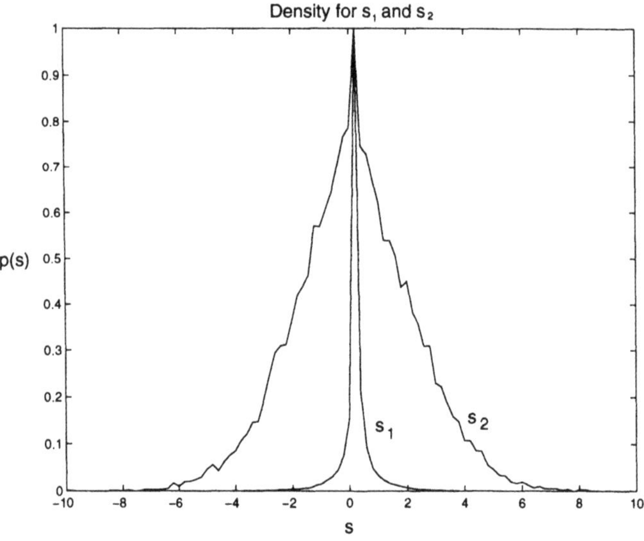

Figure 2.14. P.d.f.s of source with a low and a high kurtosis.

(kurtosis is zero). Figure 2.14 shows the density of the sources with the highest and lowest kurtosis. The kurtosis of all original signals are shown in table 2.3. The goal was to recover the sources with the different nonlinearities accounting for different

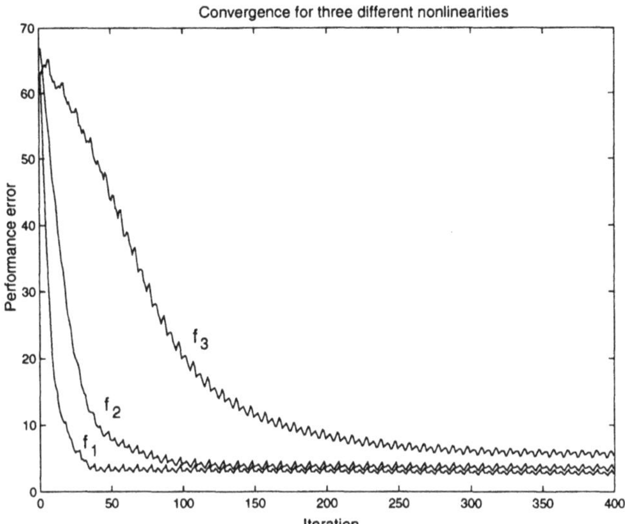

Figure 2.15. Error measure given the number of iterations. Convergence for three different nonlinearities: (a) f_1 (b) f_2 and (c) f_3.

source density estimations. The following nonlinearities were chosen

$$f_1(\mathbf{u}) = \operatorname{sign}(\mathbf{u}) \tag{2.79}$$

$$f_2(\mathbf{u}) = \tanh(\mathbf{u}) \tag{2.80}$$

$$f_3(\mathbf{u}) = \operatorname{abs}(\mathbf{u}^{0.9}) \times \operatorname{sign}(\mathbf{u}), \tag{2.81}$$

where $\times$ denotes an array multiplication. f_1 and f_3 can be derived from the generalized Gaussian nonlinearity

$$g(u_i) \propto \exp(-|u_i|^r) \tag{2.82}$$

where r is the shape parameter. For $r = 1$ eq.2.82 reduces to the Laplacian density and therefore leads to f_1. For $r = 1.9$ eq.2.82 leads to f_3. Note that for $r = 2$ eq.2.82 is the Gaussian density. Note furthermore that the relation between $f(.)$ and $g(.)$ is the derivative of the log density of the derivative of $g(.)$ and $g(.)$ approximates the c.d.f. of the source distribution. Therefore, f_1 is suited to separate Laplacian distributed data $((p(s_i) \propto \exp(-|s_i|))$ and f_3 is more likely to separate densities that are close to a Gaussian density.

In this simulation the ten sources were randomly mixed and separated with each nonlinearity f_1, f_2 and f_3. In each case the learning rate was fixed at 0.0001 and the data (20000 time points) was passed 30 times through the algorithm. Figure 2.15 shows the convergence of the error measure for all three nonlinearities. The speed of convergence for f_1 is the fastest and the convergence of f_2 is still much faster than for f_3. This result may not be surprising due to $\Delta \mathbf{W}$ which is proportional to $f(\mathbf{u})\mathbf{u}^T$ in eq.2.78. For f_1 the change in $\Delta \mathbf{W}$ is proportional to $\mathbf{u}^T$ whereas for f_2, $\Delta \mathbf{W}$ is proportional to $k\mathbf{u}^T$ where $|k| \leq 1$ depending on the slope of the nonlinearity. Effectively, the decreasing slope of the nonlinearity can be thought of as decreasing the fixed learning rate. In other words, f_1 may converge with the same speed as f_3 when the learning rate for f_1 is set much lower than for f_3.

Simular results were obtained on separating mixed sub-Gaussian distributions with the three nonlinearities. The sign of the nonlinearities f_i need to be changed according to eq.2.78 and the algorithm again converged for each nonlinearity.

The simulation results and the robustness conditions suggest that infomax effectively needs a projection into a super-Gaussian prior or a sub-Gaussian prior only. Given the general extended infomax learning rule in eq.2.78, the algorithm will converge into a separating solution.

2.9 DISCUSSIONS

2.9.1 Comparison to other algorithms and architectures

The infomax algorithm (Bell and Sejnowski, 1995) presented here for a feedforward architecture has been shown effective and efficient on several datasets. Other algorithms that were tested were slower in convergence. The same simulations were performed using the extended exploratory projection pursuit network with inhibitory lateral connections (Girolami and Fyfe, 1997b) and a hierarchical Hebbian feedforward and anti-Hebbian feedback learning network. The single feedforward neural network converged several times faster than these other architectures using the same learning rate and a block size of 1. Larger block sizes can be used in the feedforward network but not in the feedback network. The use of larger block sizes increases the convergence speed considerably due to a more reliable estimate of the switching moments k_i.

2.9.2 Applications to real world problems

The extended infomax algorithm has recently been applied to real world problems such as analyzing electroencephalographic (EEG) data (Makeig et al., 1997; Jung et al., 1998a) and functional magnetic resonance imaging (fMRI) data (McKeown et al., 1998b). Makeig et al. (1996) showed that the Bell and Sejnowski (1995) algorithm is able to linearly decompose EEG activity and artifacts. Jung et al. (1998a) show that the extended infomax algorithm is able to additionally extract sub-Gaussian artifacts such as line noise and eye movements. The reported results for the separation of eye-movement artifacts from EEG recordings have immediate application to medical and research data. Independently, Vigario et al. (1996) reported similar findings for EEG recordings using a fixed-point algorithm for ICA (Hyvaerinen and Oja, 1997a). It would be useful to compare this and other ICA algorithms on the same data sets to assess their merits. Compared to traditional techniques in EEG analysis extended infomax requires less supervision and is easy to apply (see Makeig et al. (1997); Jung et al. (1998a)). In addition to the very encouraging results on EEG data McKeown et al. (1998b) have demonstrated another successful use of the extended infomax algorithm on fMRI recordings. They investigated task-related human brain activity in fMRI data. In this application, they considered both spatial and temporal ICA and found that the extended infomax algorithm extracted sub-Gaussian temporal components that could not be extracted with the original infomax algorithm.

2.9.3 Biological plausibility

Linsker's infomax principle and the sensory coding strategy proposed by Atick (1992) were biologically motivated. However, the learning rule in eq.2.56 in a single layer feedforward neural network is non-local and would be more difficult to implement. An examples of a local learning rule using eq.2.56 is the extended exploratory projection pursuit network with inhibitory lateral connections (Girolami and Fyfe, 1997b). A feedback architecture with local learning rules is presented by Cichocki et al. (1995).

Recently Nadal and Parga (1997) have suggested that the infomax learning rule can be related to the BCM theory of synaptic plasticity (Bienenstock et al., 1982).

2.9.4 Limitations and future research

The extended infomax learning algorithm makes several assumptions that limit its effectiveness.

First, the algorithm requires the number of sensors to be the same or greater than the number of sources ($N \geq M$). The case when there are more sources than sensors, $N < M$, is of theoretical and practical interest. Given only one or two sensors that observe more than two sources is it still possible to recover all sources? Preliminary results by Lewicki and Sejnowski (1998b) suggest that an overcomplete representation of the data can to some extent extract the independent components using a priori knowledge of the source distribution. This has been applied by Lee et al. (1998c) to separate three sources from two sensors.

Second, researchers have recently tackled the problem of nonlinear mixing phenomena. Yang et al. (1997), Taleb and Jutten (1997) and Lee et al. (1997c) propose extensions when linear mixing is combined with certain nonlinear mixing models. Other approaches use self-organizing feature maps to identify nonlinear features in the data (Lin and Cowan, 1997; Pajunen and Karhunen, 1997).

Third, sources may not be stationary, i.e. sources may appear and disappear and move (speaker moving in a room). In these cases, the weight matrix $\mathbf{W}$ may change completely from one time point to the next. This is a challenging problem for all existing ICA algorithms.

Fourth, sensor noise may influence separation and should be included in the model (Nadal and Parga, 1994; Moulines et al., 1997). Much more work needs to be done to determine the effect of noise on performance.

In addition to these limitations, there are other issues that deserve further research. In particular, it remains an open question to what extent the learning rule is robust to parametric mismatch given a limited number of data points.

Despite these limitations, the extended infomax ICA algorithm presented here should have many applications where both sub-Gaussian and super-Gaussian sources need to be separated without additional prior knowledge of their statistical properties.

2.9.5 Conclusions

The extended infomax ICA algorithm is a promising generalization that satisfies a general stability criterion for mixed sub-Gaussian and super-Gaussian sources (Cardoso and Laheld, 1996). Based on the learning algorithm first derived by Girolami (1997b) and the natural gradient, the extended infomax algorithm has shown excellent performance on several large real data sets derived from electrical and blood flow measurements of functional activity in the brain. Compared to the originally proposed infomax algorithm (Bell and Sejnowski, 1995), the extended infomax algorithm separates a wider range of source signals whilst maintaining its simplicity.

Notes

1. as detailed in section 4 of Bell and Sejnowski (1995)

2. see eqs. 40 and 41 in their paper.

3. Symmetric bimodal densities considered in this paper are sub-Gaussian, however this is not always the case.

4. The presented estimation theory is related to the semiparametrical statistical approach by Amari and Cardoso (1997) and the stability analysis of adaptive blind source separation (Amari et al., 1997a)

3 A UNIFYING INFORMATION-THEORETIC FRAMEWORK FOR ICA

Good order is the foundation of all things.
Edmund Burke

3.1 OVERVIEW

This chapter shows that different theories recently proposed for ICA lead to the same iterative learning algorithm for blind separation of mixed independent sources. Those theories are reviewed and it is suggested that information theory can be used to unify several lines of research.

Bell and Sejnowski (1995) put the blind source separation problem into an information theoretic framework and demonstrated the separation and deconvolution of mixed sources. Their adaptive methods are more plausible from a neural processing perspective than the cumulant-based cost functions proposed by Comon (1994). A similar adaptive method for source separation was proposed by Cardoso and Laheld (1996). Other algorithms for performing ICA have been proposed from different viewpoints. Maximum Likelihood Estimation (MLE) approaches to ICA were first proposed by Gaeta and Lacoume (1990) and elaborated by Pham et al. (1992). Pearlmutter and Parra (1996), MacKay (1996) and Cardoso (1997) showed that the infomax approach of Bell and Sejnowski (1995) and the maximum likelihood estimation approach are equivalent. Girolami and Fyfe (1997c) motivated by information-theoretic indices for Exploratory Projection Pursuit (EPP) used marginal negentropy as a projection index and showed that kurtosis-seeking projection pursuit will extract one of the underlying sources from a linear mixture. A multiple output EPP network was developed to allow full separation of all the underlying sources (Girolami and Fyfe, 1997b). Nonlinear PCA algorithms for ICA which have been developed by Karhunen and Joutsensalo (1994), Xu (1993) and Oja (1997) can also be viewed from the infomax principle since they approximately minimize the sum of squares of the fourth-order marginal cumulants (Comon, 1994) and therefore approximately minimize the mutual information of the network outputs (Girolami and Fyfe, 1997d). Bell and Sejnowski (1995) have pointed out a similarity between their infomax algorithm and the Bussgang algorithm in signal processing and Lambert (1996) elucidated the connection between three different Bussgang cost functions. Lee et al. (1998a) show how the Bussgang property relates to the infomax principle and how all of these seemingly different approaches can be put into a unifying framework for the source separation problem based on an information theoretic approach.

This chapter is organized as follows: Section 3.2 reviews briefly the infomax approach by Bell and Sejnowski (1995). Section 3.3, 3.4, 3.5, 3.6 and 3.7 describe respectively the relation between infomax, MLE, negentropy maximization, nonlinear PCA, higher-order statistics and the Bussgang property. Finally conclusions are presented in section 3.8.

3.2 INFORMATION MAXIMIZATION

Nadal and Parga (1994) showed that in the low-noise case, the maximum of the mutual information between the inputs $\mathbf{x}$ and outputs $\mathbf{y}$ of a neural processor implied that the output distributions were factorial. In other words, maximizing the information transfer in a nonlinear neural network minimizes the mutual information among the outputs (factorial code) when optimization is performed over both the synaptic weights $\mathbf{W}$ and the nonlinear transfer function $g(\mathbf{u})$. Roth and Baram (1996) and Bell and Sejnowski (1995) independently derived stochastic gradient learning rules for this maximization and applied them, respectively to forecasting, time series analysis, and the blind separation of sources. Furthermore, Deco and Obradovic (1996) present a detailed study of an unsupervised information-theoretic approach to ICA.

Bell and Sejnowski (1995) proposed a simple learning algorithm for a feedforward neural network that blindly separates linear mixtures $\mathbf{x}$ of independent sources $\mathbf{s}$ using information maximization. They show that maximizing the joint entropy $H(\mathbf{y})$ of the output of a neural processor can approximately minimize the mutual information among the output components $y_i = g(u_i)$ where $g(u_i)$ is an invertible monotonic nonlinearity and $\mathbf{u} = \mathbf{W}\mathbf{x}$.

The joint entropy at the outputs of a neural network is

$$H(y_1, \cdots, y_N) = H(y_1) + \cdots + H(y_N) - I(y_1, \cdots, y_N), \tag{3.1}$$

where $H(y_i)$ are the marginal entropies of the outputs and $I(y_1, \cdots, y_N)$ is their mutual information. Maximizing $H(y_1, \cdots, y_N)$ consists of maximizing the marginal entropies and minimizing the mutual information. The outputs $\mathbf{y}$ are amplitude-bounded random variables and therefore the marginal entropies are maximum for a uniform distribution of y_i. Maximizing the joint entropy will also decrease $I(y_1, \cdots, y_N)$ since the mutual information is always positive. For $I(y_1, \cdots, y_N) = 0$ the joint entropy is the sum of marginal entropies

$$H(y_1, \cdots, y_N) = H(y_1) + \cdots + H(y_N). \tag{3.2}$$

The maximal value for $H(y_1, \cdots, y_N)$ is achieved when the mutual information among the bounded random variables $y_1, \cdots, y_N$ is zero and their marginal distribution is uniform. As shown below, this implies that the nonlinearity $g(u_i)$ has the form of the cumulative density function (c.d.f.) of the true source distribution s_i. Bell and Sejnowski (1995) chose the nonlinearity to be a fixed logistic function. This is equivalent to assuming a prior distribution of the sources: a super-Gaussian distribution with heavy tails and a peak centered at the mean. The weights $\mathbf{W}$ are determined by maximizing the joint entropy with respect to $\mathbf{W}$. The derivative of eq.3.1 with respect to $\mathbf{W}$ can be written in terms of the KL divergence between the multivariate uniform distribution denoted as $p_1(\mathbf{y})$ and the multivariate uniform estimate $p(\mathbf{y})$.

$$\frac{\partial H(\mathbf{y})}{\partial \mathbf{W}} = \frac{\partial}{\partial \mathbf{W}}(-D(p_1(\mathbf{y})\|p(\mathbf{y}))). \tag{3.3}$$

In the limit when the transfer function $g(u_i)$ and $\mathbf{W}$ are optimized the joint entropy $H(\mathbf{y})$ is maximum and $p(\mathbf{y}) = p_1(\mathbf{y})$ so that $I(\mathbf{y}) = 0$. If $g(u_i)$ is an invertible mapping from u_i to y_i, the KL divergence in eq.3.3 is equal to the KL divergence between the estimate of the source distribution $p(\mathbf{u})$ and the sources $p(\mathbf{s})$.

$$D(p_1(\mathbf{y})\|p(\mathbf{y})) = D(p(\mathbf{s})\|p(\mathbf{u})). \tag{3.4}$$

Since the KL divergence is invariant under an invertible transformation. If the mutual information between the outputs is zero $I(y_1, \cdots, y_N) = 0$, the mutual information before the nonlinearity $I(u_1, \cdots, u_N)$ must also be zero since the nonlinearity does not introduce any dependencies.

The learning rule with the natural gradient extension was derived in chapter 2 as

$$\Delta \mathbf{W} \propto \left[(\mathbf{W}^T)^{-1} - \varphi(\mathbf{u})\mathbf{x}^T\right], \tag{3.5}$$

where

$$\varphi(\mathbf{u}) = -\frac{\frac{\partial p(\mathbf{u})}{\partial \mathbf{u}}}{p(\mathbf{u})} = \left[-\frac{\frac{\partial p(u_1)}{\partial u_1}}{p(u_1)}, \cdots, -\frac{\frac{\partial p(u_N)}{\partial u_N}}{p(u_N)}\right]^T. \tag{3.6}$$

3.3 NEGENTROPY MAXIMIZATION

Another approach related to minimizing the mutual information between the u_i's is maximizing negentropy (Girolami, 1997b). Girolami and Fyfe (1997d, 1997b), motivated by information-theoretic indices for Exploratory Projection Pursuit (EPP) used marginal negentropy as a projection index. EPP is a statistical method that allows structure in high-dimensional data to be identified (Friedman, 1987). This is achieved by projecting the data onto a low-dimensional subspace and searching for structure in the projection. Projections that identify non-Gaussian structure such as multiple modes are interesting from the point of view of identifying potential higher-order structure within high-dimensional data. Projections that are maximally non-Gaussian are highly desirable in pursuing informative views of the data (Friedman, 1987). Girolami (1997b) showed that if the observed data fits a latent variable model (Everitt, 1984), which conforms to the deterministic ICA mixing model, then a kurtosis-seeking projection pursuit will extract one of the underlying sources. A multiple output EPP network was also developed to allow full separation of all the underlying sources (Girolami and Fyfe, 1997b). Jones and Sibson (1987) noted that approximately symmetrical and almost Gaussian (low kurtosis) clustered projections can sometimes be difficult to identify with indices based on third- and fourth-order moments and suggested the use of indices based on information theoretic criteria. Girolami and Fyfe (1997b) developed single and multiple output algorithms for EPP based on negentropy maximization. He showed that a negentropy maximizing pursuit will perform a general ICA on sources which may be either sub- or super-Gaussian. The negentropy of the output neurons can be stochastically maximized by driving their distributions maximally away from Gaussian distributions. Girolami (1997b) showed that maximizing the output data negentropy is identical to minimizing the mutual information of the output data which has been shown to be equivalent to ICA for observed data that can be modeled as a sum of independent latent variables. A brief derivation follows:

Negentropy is defined as the KL divergence between $p(\mathbf{u})$ and the Gaussian distribution $p_G(\mathbf{u})$ with the same mean and covariance as $p(\mathbf{u})$ (Cover and Thomas, 1991)

$$J(\mathbf{u}) = D(p(\mathbf{u})\|p_G(\mathbf{u})) = \int p(\mathbf{u}) \log \frac{p(\mathbf{u})}{p_G(\mathbf{u})} d\mathbf{u}, \tag{3.7}$$

where $\mathbf{u}$ is the vector of estimated sources given the parameters $\mathbf{W}$ ($\mathbf{u} = \mathbf{W}\mathbf{x}$). The parametric form of the output is factorable $\prod_{i=1}^{N} p(u_i)$ with the equality $p(\mathbf{u}) = \prod_{i=1}^{N} p(u_i)$ holding only when all u_i's are independent, i.e. the mutual information is zero ($I(\mathbf{u}) = 0$). Assume that $\mathbf{u}$ is decorrelated and that u_i's are factorable but not factorized ($J(\mathbf{u}) \neq \sum_{i=1}^{N} J(u_i)$).

$$\begin{aligned}
\sum_{i=1}^{N} J(u_i) &= \sum_{i=1}^{N} D(p(u_i)\|p_G(u_i)) && (3.8)\\
&= \int p(u_1)\log\frac{p(u_1)}{p_G(u_1)}du_1 + \cdots + \int p(u_N)\log\frac{p(u_N)}{p_G(u_N)}du_N && (3.9)\\
&= \int p(\mathbf{u})\log\frac{\prod_{i=1}^{N} p(u_i)}{\prod_{i=1}^{N} p_G(u_i)}d\mathbf{u} && (3.10)\\
&= \int p(\mathbf{u})\log\frac{\prod_{i=1}^{N} p(u_i)}{p_G(\mathbf{u})}d\mathbf{u} && (3.11)\\
&= \int p(\mathbf{u})\log\frac{\prod_{i=1}^{N} p(u_i)}{p(\mathbf{u})}d\mathbf{u} + \int p(\mathbf{u})\log\frac{p(\mathbf{u})}{p_G(\mathbf{u})}d\mathbf{u} && (3.12)\\
&= D(\prod_{i=1}^{N} p(u_i)\|p(\mathbf{u})) + J(\mathbf{u}) && (3.13)\\
&= -I(\mathbf{u}) + J(\mathbf{u}). && (3.14)
\end{aligned}$$

The sum of negentropies can be written as a sum of KL divergences. The substitution $p_G(\mathbf{u}) = \prod_{i=1}^{N} p_G(u_i)$ in eq.3.11 follows from the assumption that $\mathbf{u}$ is decorrelated. The first term in eq.3.14 is the negative of the mutual information ($-I(\mathbf{u})$). The second term can be further expanded

$$\begin{aligned}
\sum_{i=1}^{N} J(u_i) &= -I(\mathbf{u}) - H(\mathbf{u}) - \int p(\mathbf{u})\log p_G(\mathbf{u})d\mathbf{u} && (3.15)\\
&= -I(\mathbf{u}) - H(\mathbf{x}) - \log(|\det(\mathbf{W})|)\\
&\quad -\frac{1}{2}\log((2\pi e)^N \det(\langle \mathbf{u}\mathbf{u}^T\rangle)). && (3.16)
\end{aligned}$$

There are two terms that need to be justified for the equality of eq.3.15 and eq.3.16. First, the term $H(\mathbf{u})$ can be substituted by $H(\mathbf{x})+\log(|\det(\mathbf{W})|)$ because of the p.d.f. transformation equality in eq.2.26 and $\mathbf{u} = \mathbf{W}\mathbf{x}$. Second, the integral $\int p(\mathbf{u})\log p_G(\mathbf{u})d\mathbf{u}$ is the entropy of a Gaussian distribution for any distribution of $p(\mathbf{u})$ when $p(\mathbf{u})$ and $p_G(\mathbf{u})$ yield the same covariance matrix (Cover and Thomas, 1991, page 234). Since the u_i's were assumed uncorrelated its covariance matrix is identity and therefore the determinant is one. Therefore it follows that

$$\sum_{i=1}^{N} J(u_i) = -I(\mathbf{u}) - H(\mathbf{x}) - \frac{1}{2}\log((2\pi e)^N). \quad (3.17)$$

The negentropy can be maximized by using the stochastic gradient ascent

$$\frac{\partial}{\partial \mathbf{W}}\sum_{i=1}^{N} J(u_i) = \frac{\partial}{\partial \mathbf{W}}(-I(\mathbf{u}) - H(\mathbf{x}) - \frac{1}{2}\log((2\pi e)^N)). \quad (3.18)$$

The input data entropy and the nonlinear function of the input data covariance matrix are not functions of the weight parameters and so maximizing the sum of marginal negentropies with respect to $\mathbf{W}$ is equivalent to minimizing the mutual information

$$\frac{\partial}{\partial \mathbf{W}} \sum_{i=1}^{N} J(u_i) = \frac{\partial}{\partial \mathbf{W}}(-I(\mathbf{u})). \tag{3.19}$$

This leads to exactly the same learning rule as in section 3.3 using infomax. Maximizing $\sum_{i=1}^{N} J(u_i)$ with respect to $\mathbf{W}$ in eq.3.16 gives

$$\begin{aligned} \frac{\partial}{\partial \mathbf{W}} \sum_{i=1}^{N} J(u_i) &= \frac{\partial}{\partial \mathbf{W}} \left[\int p(\mathbf{u}) \log(\prod_{i=1}^{N} p(u_i)) d\mathbf{u} + \frac{1}{2} \log((2\pi e)^N) \right] \\ &= \frac{\partial}{\partial \mathbf{W}} \left[E\{\log(\prod_{i=1}^{N} p(u_i))\} \log(\det(\mathbf{W})) + \right. \\ &\quad \left. + \frac{1}{2} \log((2\pi e)^N) \right]. \end{aligned} \tag{3.20}$$

Note that as in eq.2.45 only the first and second terms in eq.3.20 depend on $\mathbf{W}$

$$\frac{\partial}{\partial \mathbf{W}} \sum_{i=1}^{N} J(u_i) = \mathbf{W}^{-T} + \left(\frac{\frac{\partial p(\mathbf{u})}{\partial \mathbf{u}}}{p(\mathbf{u})} \right) \mathbf{x}^T. \tag{3.21}$$

Although the derivation of the learning rule in eq.3.21 depends on the assumption that $\mathbf{u}$ is decorrelated Girolami (1997b) showed that a slightly different objective function related to maximizing the marginal negentropies leads to the same learning algorithm in eq.3.21 without making the assumption that $\mathbf{u}$ is decorrelated.

3.4 MAXIMUM LIKELIHOOD ESTIMATION

The goal of MLE is to model the observation $\mathbf{x}$ as being generated from latent variables $\mathbf{s}$ via a linear mapping $\mathbf{A}$. In the noiseless case, a parametric density estimator $\hat{p}(\mathbf{x};\mathbf{a})$ can be used to find the parameter vector $\mathbf{a}$ that minimizes the difference between the generative model $\hat{p}(\mathbf{x};\mathbf{a})$ and the observed distribution $p(\mathbf{x})$. Note that $\mathbf{a}$ can be considered the basis vectors of $\mathbf{A}$ so that $\hat{p}(\mathbf{x};\mathbf{a})$ is an estimate of the observed vector $p(\mathbf{x})$. The difference between the estimate and the observation can be measured using the KL divergence

$$D(p(\mathbf{x}), \hat{p}(\mathbf{x};\mathbf{a})) = \int p(\mathbf{x}) \log \frac{p(\mathbf{x})}{\hat{p}(\mathbf{x};\mathbf{a})} d\mathbf{x} = H(\mathbf{x}) - \int p(\mathbf{x}) \log \hat{p}(\mathbf{x};\mathbf{a}), \tag{3.22}$$

where $p(\mathbf{x})$ is the p.d.f. of the observation $\mathbf{x}$ and $\hat{p}(\mathbf{x};\mathbf{a})$ is a parametric estimate of the distribution $p(\mathbf{x})$. The divergence $D(p(\mathbf{x})\|\hat{p}(\mathbf{x};\mathbf{a}))$ is zero only if our estimate

$\hat{p}(\mathbf{x};\mathbf{a})$ matches the observation $p(\mathbf{x})$. Pearlmutter and Parra (1997) and Cardoso (1997) showed that infomax and MLE are equivalent for ICA, as briefly described here. The normalized log-likelihood of $\hat{p}(\mathbf{x};\mathbf{a})$ is

$$L(\mathbf{a}) = \frac{1}{N}\sum_{i=1}^{N} \log \hat{p}(\mathbf{x}_i;\mathbf{a}), \tag{3.23}$$

where N is the number of independent realizations of $\mathbf{x}$. The log-likelihood converges in probability, by the law of large numbers, to its expectation

$$L(\mathbf{a}) = \int p(\mathbf{x}) \log \hat{p}(\mathbf{x};\mathbf{a}) d\mathbf{x}. \tag{3.24}$$

Note that this can be rewritten

$$\begin{aligned} L(\mathbf{a}) &= -\int p(\mathbf{x}) \log p(\mathbf{x}) d\mathbf{x} - \int p(\mathbf{x}) \log \frac{p(\mathbf{x})}{\hat{p}(\mathbf{x};\mathbf{a})} d\mathbf{x} \\ &= H(\mathbf{x}) - D(p(\mathbf{x})\|\hat{p}(\mathbf{x};\mathbf{a})). \end{aligned} \tag{3.25}$$

Since $H(\mathbf{x})$ is not dependent on $\mathbf{W}$, maximizing the log-likelihood minimizes the KL divergence between the observed density $p(\mathbf{x})$ and the estimated density $\hat{p}(\mathbf{x};\mathbf{a})$

$$\frac{\partial L(\mathbf{a})}{\partial \mathbf{W}} = -\frac{\partial}{\partial \mathbf{W}} D(p(\mathbf{x})\|\hat{p}(\mathbf{x};\mathbf{a})). \tag{3.26}$$

Since $\mathbf{A}$ is an invertible matrix and the KL divergence is invariant under an invertible transformation, minimizing the KL divergence in eq.3.26 minimizes the KL divergence between the estimate of the sources $p(\mathbf{u})$ and the true source distribution $p(\mathbf{s})$

$$\frac{\partial L(\mathbf{a})}{\partial \mathbf{W}} = -\frac{\partial}{\partial \mathbf{W}} D(p(\mathbf{s})\|\hat{p}(\mathbf{u})). \tag{3.27}$$

Therefore eq.3.27 and eq.3.3 are equivalent for ICA.

3.5 HIGHER-ORDER MOMENTS AND CUMULANTS

In the previous sections, the nonlinearity of the output approximated the c.d.f. of the true source density. Here cumulants are examined to study the higher-order correlations between the sources.

If the observed vector has a covariance matrix $\langle \mathbf{x}\mathbf{x}^T \rangle = E\{\mathbf{x}\mathbf{x}^T\}$ then the mutual information in eq.2.4 can be expressed as (Comon, 1994)

$$I(\mathbf{x}) = J(\mathbf{x}) - \sum_{i=1}^{N} J(x_i) + \frac{1}{2} \log \frac{(\prod_{i=1}^{N} \langle x_i^2 \rangle)}{\det(\langle \mathbf{x}\mathbf{x}^T \rangle)}, \tag{3.28}$$

where $\langle x_i^2 \rangle$ in eq.3.28 are the diagonal elements of the covariance matrix. $J(\mathbf{x})$ is the multivariate negentropy as in eq.3.7 and $J(x_i)$ are the marginal negentropies

$$J(x_i) = \int p(x_i) \log \frac{p(x_i)}{p_G(x_i)} dx_i. \tag{3.29}$$

If a spatial whitening transformation (diagonalization of the covariance matrix) is used to remove the second-order redundancy in the data, $\tilde{\mathbf{x}} = \mathbf{V}\mathbf{x}$, where $\mathbf{V}$ denotes the whitening transformation matrix and $\langle \tilde{\mathbf{x}}\tilde{\mathbf{x}}^T \rangle = \mathbf{I}$ then $\det(\langle \tilde{\mathbf{x}}\tilde{\mathbf{x}}^T \rangle) = 1$ and the mutual information of the spatially white data reduces to

$$I(\tilde{\mathbf{x}}) = J(\tilde{\mathbf{x}}) - \sum_{i=1}^{N} J_i(\tilde{x}_i). \tag{3.30}$$

A further transformation $\mathbf{u} = \mathbf{W}\tilde{\mathbf{x}}$ using higher-order correlations is required to reduce the remaining redundancy within the vector for non-Gaussian sources. This transformation seeks an orthogonal matrix that accounts for the correct rotation of the data. Comon (1994) minimized the degree of dependence among outputs using contrast functions approximated by the Edgeworth expansion of the KL divergence. He determined the orthogonal matrix from the higher-order cumulants. Note that cumulants are used to describe characteristics of non-Gaussian processes. The truncated Edgeworth expansion (Stuart and Ord, 1987) of $p(u_i)$ written in terms of its nth-order cumulants and Hermite polynomials, denoted as k_n and h_n respectively, is

$$\begin{aligned} p(u_i) = p_G(u_i) \Bigg[1 &+ \frac{1}{3!}k_3 h_3(u_i) + \frac{1}{4!}k_4 h_4(u_i) + \frac{10}{6!}k_4^2 h_6(u_i) \\ &+ \frac{1}{5!}k_5 h_5(u_i) + \frac{35}{7!}k_3 k_4 h_7(u_i) + \frac{280}{9!}k_3^3 h_9(u_i) \\ &+ \frac{1}{5!}k_5 h_5(u_i) + \frac{56}{8!}k_3 k_5 h_8(u_i) + \frac{35}{8!}k_4^2 h_8(u_i) \\ &+ \frac{2100}{10!}k_3^2 k_4 h_{10}(u_i) + \frac{15400}{12!}k_3^4 h_{12}(u_i) \Bigg], \end{aligned} \tag{3.31}$$

where $p_G(u_i)$ denotes the Gaussian density. The cumulants k_n are coefficients related to the form of the p.d.f. of u_i and they can be expressed in terms of moments. The terms $h_k(u_i)$ are the orthogonal Hermite polynomials defined as (Stuart and Ord, 1987)

$$(-1)^k \frac{\partial^k p_G(u_i)}{\partial u^k} = h_k(u_i) p_G(u_i), \tag{3.32}$$

which can be computed recursively

$$\begin{aligned} h_0(u_i) &= 1 \\ h_k(u_i) &= u_i h_{k-1} - (k-1) h_{k-2}. \end{aligned} \tag{3.33}$$

The validity of the truncated series expansion approximation in eq.3.31 is discussed in Stuart and Ord (1987). Expansion terms higher than fourth-order can lead to excessive fluctuations at the tails of the distribution leading potentially to negative values. Therefore, the expansion in eq.3.31 is truncated at fourth-order. After substituting the expression for marginal negentropies $J(u_i)$ in eq.3.29 into eq.3.31 (Comon, 1994), $J(u_i)$ becomes

$$J(u_i) \cong \frac{1}{12}k_3^2(i) + \frac{1}{48}k_4^2(i) + \frac{7}{48}k_3^4(i) + \frac{1}{8}k_3^2(i)k_4(i). \tag{3.34}$$

Assuming that the p.d.f. of the signals under consideration are approximately symmetric then the third-order cumulants will have a negligible contribution in eq.3.34. The mutual information in eq.3.28 of the transformed data $\mathbf{u}$ is now approximated by

$$I(\mathbf{u}) \cong J(\mathbf{u}) - \frac{1}{48}\sum_{i=1}^{N} k_4^2(i). \tag{3.35}$$

$J(\mathbf{u})$ is invariant under an orthogonal transformation

$$\begin{aligned} J(\mathbf{u}) &= \int p(\mathbf{u}) \log \frac{p(\mathbf{u})}{p_G(\mathbf{u})} d\mathbf{u} \\ &= H(\mathbf{u}) - \frac{1}{2}\log((2\pi e)^N \det(\langle \mathbf{u}\mathbf{u}^T \rangle)) \\ &= H(\tilde{\mathbf{x}}) + \log(\det(\mathbf{W})) - \frac{1}{2}\log((2\pi e)^N \det(\mathbf{W}\langle \tilde{\mathbf{x}}\tilde{\mathbf{x}}^T \rangle \mathbf{W}^T)) \\ &= H(\tilde{\mathbf{x}}) - \frac{1}{2}\log((2\pi e)^N \det(\langle \tilde{\mathbf{x}}\tilde{\mathbf{x}}^T \rangle)) \\ &= H(\tilde{\mathbf{x}}) - H_G(\tilde{\mathbf{x}}) = J(\tilde{\mathbf{x}}), \end{aligned} \tag{3.36}$$

where $H_G(\tilde{\mathbf{x}})$ is the entropy of a normal density the following matrix determinant equalities have been employed

$$\det(\mathbf{W}\langle \tilde{\mathbf{x}}\tilde{\mathbf{x}}^T \rangle \mathbf{W}^T) = \det(\mathbf{W})\det(\langle \tilde{\mathbf{x}}\tilde{\mathbf{x}}^T \rangle)\det(\mathbf{W}^T) \tag{3.37}$$

$$\det(\mathbf{W}^T) = \det(\mathbf{W}). \tag{3.38}$$

Since $\mathbf{u}$ is a result of a rotation of $\tilde{\mathbf{x}}$ the negentropy $J(\mathbf{u})$ is equal to $J(\tilde{\mathbf{x}})$ and the approximation for mutual information can be rewritten as

$$I(\mathbf{u}) \cong J(\tilde{\mathbf{x}}) - \frac{1}{48}\sum_{i=1}^{N} k_4^2(i). \tag{3.39}$$

Thus, under an orthogonal transformation, the mutual information of the data can be approximately minimized by maximizing the sum of squares of the fourth-order marginal cumulants. Maximizing the contrast function is approximately equivalent to maximizing the sum of marginal negentropies. This corroborates the claim that maximizing the marginal negentropies with respect to $\mathbf{W}$ minimizes mutual information.

$$\frac{\partial}{\partial \mathbf{W}} I(\mathbf{u}) \cong \frac{\partial}{\partial \mathbf{W}}(-\sum_{i=1}^{N} k_4^2(i)). \tag{3.40}$$

Therefore, Comon (1994) proposed the following contrast function

$$\Phi_{\max} = \sum_{i=1}^{N} k_4^2(i). \tag{3.41}$$

Here, the higher-order statistics are approximated by cumulants up to 4th-order and their maximization requires intensive computation using a batch-based method.

3.6 NONLINEAR PCA

The nonlinear extension of Oja's Principle Component Analysis (PCA) subspace network (Oja, 1982), originally developed by Karhunen and Joutsensalo (1994) and Xu (1993), has no an apparent connection to the infomax principle, but has been shown to separate whitened linear mixtures of sources (Karhunen et al., 1997c; Oja and Karhunen, 1995; Karhunen et al., 1995). A major shortcoming of the algorithm is that is has been restricted to the separation of sub-Gaussian sources, because of stability requirements. Another property is that the data have to be prewhitened. Those two characteristics have led Girolami and Fyfe (1997d) to relate the nonlinear PCA algorithm to the infomax principle showing that it is an approximate online adaptive equivalent of the batch algorithm proposed by Comon (1994).

In this section, the results in Girolami and Fyfe (1997d) and their generalization to cope with sub- and super-Gaussian source distributions are summarized. Their generalization is an alternative form of the nonlinear PCA rule which satisfies the dynamic and asymptotic stability criteria for the algorithm (Girolami and Fyfe, 1997d).

In nonlinear PCA, the input signals $\mathbf{x}$ are first prewhitened giving $\tilde{\mathbf{x}}$, where $\langle\tilde{\mathbf{x}}\tilde{\mathbf{x}}^T\rangle = \mathbf{I}$. The learning rule is an approximate stochastic gradient descent algorithm that minimizes the mean-squared error incurred in representing a vector by a nonlinear projection $f(\mathbf{W}\tilde{\mathbf{x}})$ onto a basis of reduced dimensionality

$$\tilde{\mathbf{x}} = \tilde{\mathbf{x}}' + e = \mathbf{W}f(\mathbf{W}\tilde{\mathbf{x}}) + e, \tag{3.42}$$

where $\tilde{\mathbf{x}}$ is a nonlinear estimate of $\tilde{\mathbf{x}}$ and e denotes the estimation error. Next, minimize a cost function $C(\mathbf{W})$ to find a linear transformation $\mathbf{W}$ giving $\mathbf{u} = \mathbf{W}\tilde{\mathbf{x}}$ where $\mathbf{u}$ are the estimated sources and $\mathbf{W}$ is constrained to be orthonormal $\mathbf{W}^T\mathbf{W} = \mathbf{I}$.

$$C(\mathbf{W}) = \mathbf{1}^T E\{(\tilde{\mathbf{x}} - \mathbf{W}f(\mathbf{W}\tilde{\mathbf{x}}))^2\}, \tag{3.43}$$

where $C(\mathbf{W})$ is a scalar resulting from an inner product and $\mathbf{1}^T$ is a row vector of length N with ones as its elements. Rewriting eq.3.43 in its transpose form gives

$$C(\mathbf{W}) = E\{(\tilde{\mathbf{x}}^T\tilde{\mathbf{x}} - f^T(\mathbf{W}\tilde{\mathbf{x}}) - \tilde{\mathbf{x}}^T\mathbf{W}f(\mathbf{W}\tilde{\mathbf{x}}) + f^T(\mathbf{W}\tilde{\mathbf{x}})\mathbf{W}^T\mathbf{W}f(\mathbf{W}\tilde{\mathbf{x}}))\}. \tag{3.44}$$

Since the observed data is spatially white it follows that: $E\{\mathbf{W}^T\tilde{\mathbf{x}}\tilde{\mathbf{x}}^T\mathbf{W}\} = \mathbf{I}$. Assuming unit variance for the independent components u_i the cost function is now

$$C(\mathbf{W}) = N + E\{(-f^T(\mathbf{u})\mathbf{u} - \mathbf{u}^T f(\mathbf{u}) + f^T(\mathbf{u})f(\mathbf{u}))\}. \tag{3.45}$$

For a polynomial such as $(f(\mathbf{u}) = \frac{\mathbf{u}^3}{3})$ or $(f(\mathbf{u}) = -\frac{\mathbf{u}^3}{3})$ or a hyperbolic nonlinear function which has a cubic as the dominating element, the term is $f(\mathbf{u}) = \frac{\mathbf{u}^3}{3}$ and therefore

$$C(\mathbf{W}) \cong N + E\{-(\frac{\mathbf{u}^3}{3})^T\mathbf{u} - \mathbf{u}^T\frac{\mathbf{u}^3}{3} + (\frac{\mathbf{u}^3}{3})^T\frac{\mathbf{u}^3}{3}\}. \tag{3.46}$$

Now the rightmost term $\frac{(\mathbf{u}^3)^T\mathbf{u}^3}{9}$ can be neglected as $\frac{2}{3}\mathbf{u}^T\mathbf{u}^3 >> \frac{(\mathbf{u}^3)^T\mathbf{u}^3}{9}$ is satisfied for white standardized data (Girolami, 1997b). The cost function can be rewritten as

$$C(\mathbf{W}) \cong N - \frac{2}{3}E\{\sum_{i=1}^{N} u_i^4\} = (N-2) - 2/3(E\{\sum_{i=1}^{N} u_i^4\} - 3), \qquad (3.47)$$

where N is the number of sources and the term $(E\{u_i^4\} - 3)$ is the expression for the fourth-order marginal cumulant (unnormalized kurtosis). Hence, for spatially white standardized data the cost function can be considered as the negative sum of the marginal fourth order cumulants of the linearly transformed data

$$C(\mathbf{W}) \cong -\sum_{i=1}^{N} k_4(i). \qquad (3.48)$$

Minimizing the cost function in eq.3.48 is equivalent to maximizing the sum of fourth-order cumulants when the kurtosis of the estimated sources is positive (super-Gaussian). Optimization of eq.3.48 with respect to $\mathbf{W}$ is equivalent to maximization of the sum of squares of the marginal fourth-order cumulants, for mixtures of strictly super-Gaussian sources. The function

$$\Phi_{max} = \sum_{i=1}^{N} k_4^2(i), \qquad (3.49)$$

is equivalent to Comon's contrast function in eq.3.41. Comon (1994) has shown that maximizing this contrast function approximately minimizes the mutual information.

Consider now the case when the activation function is $f(\mathbf{u}) = -\frac{\mathbf{u}^3}{3}$. Applying the same reduction as above, the cost function has the following form

$$C(\mathbf{W}) \cong \sum_{i=1}^{M} k_4(i). \qquad (3.50)$$

Minimizing the cost function in eq.3.50 is equivalent to maximizing the sum of fourth-order cumulants when the kurtosis of the estimated sources is negative (sub-Gaussian). Hence, the contrast function is the same as in eq.3.49 but for a different nonlinear term. The negatively cubic term can be understood as accounting for a different prior on the source distribution. The differences in the learning rules in eq.3.48 and eq.3.50 can be summarized and formulated in a general cost function (Girolami and Fyfe, 1997d)

$$C(\mathbf{W}) \equiv -\text{sign}(f(\mathbf{u})) \sum_{i=1}^{M} k_4(i), \qquad (3.51)$$

where $\text{sign}(f(\mathbf{u}))$ is the sign function of the nonlinearity used at the output neurons and $f(\mathbf{u}) = \pm\frac{\mathbf{u}^3}{3}$. Note that this new form of minimization of the signal representation error criterion is valid for observed data which is zero-mean and spatially white.

The MSE (Mean-Squared-Error) of the cost function in eq.3.51 relates to the mutual information as shown in section 3.5 under the further assumption that probability densities are more or less symmetric so that the third-order cumulant terms within expansion can be removed from the fourth-order approximation of the Edgeworth expansion. The mutual information can then be approximated as follows (see section 3.5)

$$I(\mathbf{u}) \cong J(\tilde{\mathbf{x}}) - \frac{1}{48}\sum_{i=1}^{N} k_4^2(i). \tag{3.52}$$

As in section 3.5 maximizing the marginal negentropies with respect to $\mathbf{W}$ minimizes mutual information giving

$$\frac{\partial}{\partial \mathbf{W}} I(\mathbf{u}) \cong \frac{\partial}{\partial \mathbf{W}}(-\sum_{i=1}^{N} k_4^2(i)), \tag{3.53}$$

which corroborates that maximizing the sum of marginal cumulants or minimizing the MSE of the cost function derived for nonlinear PCA can be interpreted as an approximate information-theoretic contrast for ICA.

3.7 BUSSGANG ALGORITHMS

Bussgang algorithms have been introduced by Bellini (1994) to perform blind deconvolution. Lambert (1996) proposes three different multichannel blind deconvolution (separation and deconvolution) algorithms based on three classes of Bussgang cost functions. These algorithms are similar to the information-theoretic learning algorithm (Bell and Sejnowski, 1995) but the relationship to the infomax algorithm is not obvious. Intuitive explanations have been proposed by Bell and Sejnowski (1995); Lambert and Bell (1997); Girolami and Fyfe (1997c) and Lee et al. (1997b). Here, it is shown how the Bussgang algorithm can be interpreted as an information-theoretic cost function.

A white zero-mean stochastic process u_t has the Bussgang property if it satisfies (Bellini, 1994)

$$E\{u_t u_{t+k}\} = E\{f(u_t) u_{t+k}\}, \tag{3.54}$$

where the subscript denotes time-points t and its time-shifted version $t+k$ and the Bussgang nonlinearity $f(.)$ is some monotone nonlinear function. The Bussgang property in eq.3.54 states that the autocorrelation function of u_t is equal to the cross-correlation function between the process u_t and the output of a nonlinearity $f(u_t)$ where both correlation functions are measured for the same lag.

The Bussgang property in eq.3.54 may be rewritten for spatial processes as follows

$$E\{u_i u_j\} = E\{f(u_i) u_j\}, \tag{3.55}$$

where the subscripts i, j denote independent (white) stochastic processes. In fact eq.3.54 differs from eq.3.55 only insofar as the subscripts refer to spatial rather than temporal samples, which allows us to relate the Bussgang property to the spatial ICA formulation. Now, the left side of eq.3.55 describes the second-order cross-correlation between two estimated sources and the right side of eq.3.55 accounts for higher-order cross-correlation between these estimates due to the nonlinearity $f(.)$ that can be thought of as a combination of higher-order terms in a Taylor series expansion.

A common way to derive a learning rule in blind deconvolution is to estimate the mean-squared error between the estimate u_i and the true source s_i. However, since the true source is not available another estimator is needed. A valid estimator would be a nonlinear estimate $f(u_i)$ where the form of the function $f(.)$ has to reflect some information about the true signals s_i. Define a cost function C that minimizes the MSE between the source estimate u_i and a Bussgang nonlinear estimate $f(u_i)$. For simplicity, consider only one source estimate u_i

$$C = E\{(u_i - f(u_i))^2\}. \tag{3.56}$$

The form of the Bussgang nonlinearity can be derived from the maximum a posteriori (MAP) model by forming a conditional log-likelihood model given the observed data as follows.

For an independent source s_i, the estimated source u_i can be modeled as the source s_i plus an independent noise source n such that $u_i = s_i + n$. Define an error variable z_i as the difference between the true source signal and the estimated source signal

$$z_i = s_i - u_i. \tag{3.57}$$

Assume that u_i can be estimated by the nonlinear function $f(u_i)$ giving

$$z_i = s_i - f(u_i). \tag{3.58}$$

The conditional density of the source given the variable z_i can be described by the MAP model

$$p(s_i|z_i) = p(z_i|s_i)p(s_i). \tag{3.59}$$

Assume that $p(z_i|s_i)$ can be modeled as a white zero-mean Gaussian process giving

$$p(z_i|s_i) = K \exp\left(-\frac{(z_i - s_i)^2}{2\sigma_u^2}\right), \tag{3.60}$$

where K is a constant and $\sigma_u^2 = \sigma_s^2 + \sigma_n^2$ is the variance of u_i. The justification of a Gaussian process for the conditional estimator $p(z_i|s_i)$ is that the sum of N ($N >> 1$) zero-mean independent sources s_i sum up to a Gaussian observation due to the central limit theorem. Substituting eq.3.60 in eq.3.59 and taking the logarithm of the conditional estimate in eq.3.59 it follows that

$$\log(p(s_i|z_i)) = \log(K) - \frac{(z_i - s_i)^2}{2\sigma_u^2} + \log(p(s_i)). \tag{3.61}$$

The derivative of eq.3.61 with respect to s_i gives

$$\frac{\partial \log(p(s_i|z_i))}{\partial s_i} = \frac{(z_i - s_i)}{\sigma_u^2} + \frac{\partial \log(p(s_i))}{\partial s_i}. \tag{3.62}$$

When the estimation error is minimized eq.3.62 is zero and solving for z_i gives the following expression

$$z_i = s_i - \sigma_u^2 \frac{\partial \log(p(s_i))}{\partial s_i}. \tag{3.63}$$

Now comparing eq.3.63 with eq.3.58 and assuming unit variance for u_i ($\sigma_u^2 = 1$), the form for the Bussgang nonlinear estimator must satisfy

$$f(s_i) = \frac{\frac{\partial p(s_i)}{\partial s_i}}{p(s_i)}, \tag{3.64}$$

which is proportional to the derivative of the log-density of the true source distribution.

Applying eq.3.64 to the initial Bussgang property by rewriting eq.3.55 in matrix form gives

$$\begin{aligned} E\{\mathbf{u}\mathbf{u}^T\} &= E\{f(\mathbf{u})\mathbf{u}^T\} \\ E\{\mathbf{u}\mathbf{u}^T\} - E\{f(\mathbf{u})\mathbf{u}^T\} &= \mathbf{0} \\ E\{\mathbf{WAss}^T\mathbf{A}^T\mathbf{W}^T\} - E\{f(\mathbf{u})\mathbf{u}^T\} &= \mathbf{0}. \end{aligned} \tag{3.65}$$

The left side of eq.3.65 is the identity matrix when $\mathbf{W} = \mathbf{A}^{-1}$ is assumed. Multiplying eq.3.65 with $\mathbf{W}$ gives

$$\left[\mathbf{I} - E\{f(\mathbf{u})\mathbf{u}^T\}\right]\mathbf{W} = \mathbf{0}. \tag{3.66}$$

The optimal Bussgang nonlinearity $f(\mathbf{u})$ when applied to ICA must be equivalent to

$$f(\mathbf{u}) = -\frac{\frac{\partial p(\mathbf{u})}{\partial \mathbf{u}}}{p(\mathbf{u})}, \tag{3.67}$$

which is precisely the score function $\varphi(\mathbf{u})$ in eq.2.44. Therefore,

$$\left[\mathbf{I} - E\{\varphi(\mathbf{u})\mathbf{u}^T\},\right]\mathbf{W} = \mathbf{0} \tag{3.68}$$

which is exactly the convergence criterion for the infomax learning rule in eq.2.45. The justification of the Bussgang nonlinearity in eq.3.64 also corroborates the infomax principle and its application to blind source separation and blind deconvolution.

3.8 CONCLUSION

Several lines of research on ICA were unified within an information-theoretic framework. This framework may be is well suited to further investigate ICA from many different theoretical viewpoints.

4 BLIND SEPARATION OF TIME-DELAYED AND CONVOLVED SOURCES

We know that this is our son, and that he was born blind.
But by what means he now seeth, we know not
or who hath opened his eyes, we know not ...
John (9:20)

4.1 OVERVIEW

This chapter considers the multichannel blind deconvolution problem. Blind deconvolution refers to the problem of determining the impulse response of a system where the output is usually accessible and the system as well as the input are inaccessible. Multichannel blind deconvolution refers to the fact that multiple channels are observable and multiple sources are mixed and convolved simultaneously.

The single channel blind deconvolution problem has been well studied in the signal processing community where algorithms have been proposed for problems such as reverberation cancelation, seismic deconvolution, and image restoration (Widrow and Stearns, 1985; Haykin, 1994a; Mendel, 1990). Many second-order decorrelation methods were proposed (Feder et al., 1993; Haykin, 1991) which give satisfying solutions when the convolving system was minimum-phase (Haykin, 1991). Usually non-minimum phase systems are assumed and blind deconvolution involves the use of nonlinear adaptive filtering algorithms designed to extract higher-order statistical information from the received signals. One class of nonlinear blind deconvolution algorithm is referred to as *Bussgang* [1] algorithm (Bellini, 1994). A more direct approach to include higher-order statistics is the use of cumulants up to fourth-order (Yellin and Weinstein, 1994; Nguyen-Thi and Jutten, 1995; Comon, 1996; Icart and Gautier, 1996).

An information-theoretic approach to blind deconvolution was proposed by Bell and Sejnowski (1995). The link between the Bussgang algorithm and the infomax algorithm is at a first glance not obvious. Bell and Sejnowski (1995) pointed out the similarity between the two algorithms and Lambert (1996) gave an intuitive explanation for the relation. In chapter 3, the Bussgang algorithm was related to infomax. Other neural multichannel blind deconvolution algorithms were proposed by Back (1994); Girolami and Fyfe (1997a); Cichocki et al. (1997); Douglas et al. (1997); Lee et al. (1997a).

The goal in this chapter is to tackle the problem of separating voices recorded in real environments. This problem is related to the *cocktail party problem* where a listener can extract one voice from an ensemble of different voices corrupted by music and noise in the background. The situation may be modeled as a linear mixing and filtering of independent sources. For this assumption, a matrix of filters must be learned that approximately inverts the mixing. Bell and Sejnowski (1995) showed that infomax can be used to deconvolve independent sources. Torkkola (1996a) extended this approach to a feedback inverting system with only cross filters. A full filter feedback system was presented in Lee et al. (1997a). Independently, Cichocki et al. (1997) derived learning rules for the full feedback system. Here, two different inverting system architectures are proposed: (1) a feedback architecture where the inverting system is approximated by a full matrix of IIR filters and (2) a feedforward architecture in which a full matrix of FIR filters is used. The advantage of using a feedback system is that a parsimony of parameters may be sufficient to approximate the inverse system. Simulations demonstrate that the algorithm is capable of finding the correct parameters and therefore is able to invert the mixing system. A disadvantage is that the full filter architectures whiten the original sources and one possible way to prevent this is to approximate the inverse system with only cross filters as proposed by Torkkola (1996b). Another aspect is the problem of finding correct time-delays for natural signals since they are correlated over time. An inaccurate time-delay estimation may result into an incorrect system inverse and therefore the time-delay learning rule for the feedforward system was omitted and the time-delays were incorporated simply as part of the FIR filter. Also, since feedback systems are limited to minimum-phase systems a feedforward system is proposed to give a more general inverse system. The learning rules for the feedforward architecture can be derived in the same manner. However, a more efficient way of computing the system inverse is to make use of the polynomial filter algebra as proposed by Lambert (1996). The learning rule is effectively updated in the frequency domain where the convolution operator becomes a simple multiplication. The FIR polynomials in the matrix notation can be treated as coefficients of a scalar matrix. The power of this method is demonstrated by separating two voices recorded in a real environment. Preliminary results of speech enhancement in an automatic speech recognition system are shown indicating that this method may be used as a preprocessing step.

Another way of separating mixed sources recorded in real environments is the time-delayed decorrelation (TDD) approach (Molgedey and Schuster, 1994; Belouchrani et al., 1997). Although the method is based on decorrelation only it achieve source separation by simultaneously decorrelating the signals for different time-lags. This approach requires non-white spectra and its performance is dependent on the degree of spectral overlap between the signals. The TDD algorithm can be extended to multichannel deconvolution (Ehlers and Schuster, 1997; Murata et al., 1998; Lee et al., 1998e) by applying the algorithm to each frequency bin in the spectrogram of the observations. Since the decorrelation aspect in TDD has its weakness it can be replaced by an ICA assumption resulting in better separation quality.

This chapter is organized as follows: The problem statement and the assumptions are formulated in section 4.2. Two architectures are proposed to solve the problem: (1) A feedback architecture where the inverting system is approximated by a full matrix of IIR filters in section 4.3 and (2) A feedforward architecture in which a full matrix of FIR filters is used (section 4.4). In each subsection, the learning rules are derived using the infomax principle and the convergence of the learning algorithms are demonstrated in simulations. The polynomial FIR matrix algebra is explained in subsection 4.4.1 and section 4.5 shows experimental results with recordings in a real environment. Section 4.6 summarizes how the infomax algorithm for multichannel source separation problem can be viewed from three classes of direct Bussgang cost functions Lambert (1996). Section 4.7 presents the time-delayed decorrelation method for the multichannel source separation problem as an alternative. This principle can be extended to to time-delayed ICA in section 4.8. Conclusions are presented in section 4.9 with a summary of this chapter and a discussion on future research issues.

4.2 PROBLEM STATEMENT AND ASSUMPTIONS

Assume that there is an N dimensional zero mean vector $\mathbf{s}(t)$ such that $\mathbf{s}(t) = [s_1(t), \cdots, s_N(t)]^T$, the components are mutually independent. The vector $\mathbf{s}(t)$ corresponds to N independent scalar valued source signals $s_i(t)$. The N signals are transmitted through a medium so that an array of N sensors picks up a set of signals $\mathbf{x}(t) = [x_1(t) \dots x_N(t)]^T$, each of which has been mixed, delayed and filtered as follows

$$x_i(t) = \sum_{j=1}^{N} \sum_{k=0}^{M-1} a_{ijk} s_j(t - D_{ij} - k). \tag{4.1}$$

D_{ij} are entries in a matrix of delays and there is an M-point filter, $\mathbf{a}_{ij}$, between the the jth source and the ith sensor. The problem is to invert this environmental scrambling without knowledge of it, thus recovering the original signals, $\mathbf{s}(t)$.

The success for separating sources depends on the assumptions made in chapter 2. In addition, there are some modifications necessary:

1. The matrix of filters is invertible.
2. Each source is white, i.e.: there are no dependencies between time points.

Assumption (1) is an extension to the requirement of a full rank matrix for instantaneous mixing. Here, it is required that the matrix of filters is full rank and therefore invertible. Assumption (2), on the other hand, is not true for natural signals. The proposed algorithm will whiten, i.e. it will remove dependencies across time which already existed in the original source signals, s_i. However, it is possible to restore the characteristic autocorrelations (amplitude spectra) of the sources by post-processing (Haykin, 1991).

The mixing process can be formulated in the frequency domain as

$$\mathbf{X}(z) = \mathbf{A}(z)\mathbf{S}(z), \tag{4.2}$$

where $\mathbf{A}(z)$ is the matrix of finite impulse response (FIR) filters and $\mathbf{S}(z)$ and $\mathbf{X}(z)$ are the vectors of source signals and mixed signals respectively. The convolution operation in the time domain corresponds to a multiplication in the frequency domain.

The true inverse is the FIR matrix inverse with elements (for a two by two case $\mathbf{W}(z)$)

$$\mathbf{W}(z) = \begin{bmatrix} W_{11}(z) & W_{12}(z) \\ W_{21}(z) & W_{22}(z) \end{bmatrix} = \begin{bmatrix} \frac{A_{22}(z)}{\Delta(z)} & \frac{-A_{21}(z)}{\Delta(z)} \\ \frac{-A_{12}(z)}{\Delta(z)} & \frac{A_{11}(z)}{\Delta(z)} \end{bmatrix}, \tag{4.3}$$

where $\Delta(z)$ is the determinant of the filter matrix

$$\Delta(z) = A_{11}(z)A_{22}(z) - A_{12}(z)A_{21}(z). \tag{4.4}$$

In general the elements of the inverse system W_{ij} can be approximated by FIR filters. Feedback inverse systems may be used as well which offer a more compact representation of the solution in terms of the number of filter-taps required for a sufficient approximation. The feedback system can be realized by Infinite Impulse Response (IIR) filters and is also called an Auto Regressive Moving Average (ARMA) model [2]. One major disadvantage of the ARMA model is that it is applicable to the inversion of minimum-phase systems only, i.e., the transfer function of the system has all of its poles and zeros inside the unit circle in the z-plane. Then the unmixing system will be stable.

The following sections consider the learning rules and give simulation results for both systems: feedback and feedforward.

4.3 FEEDBACK ARCHITECTURE

Torkkola (1996a) has addressed the problem of solving the delay-compensation problem with a feedback architecture. Such an architecture can, in principle, solve this problem, as shown earlier by Platt and Faggin (1992). Torkkola (1996b) also generalized the feedback architecture to remove dependencies across time, to achieve the deconvolution of mixtures which have been filtered, as in eq.4.1. His architecture for doing this can be expressed as

$$u_i(t) = w_i x_i(t) - \sum_{\substack{j=1 \\ j\neq i}}^{N} \sum_{k=1}^{M} w_{ijk} u_j(t - d_{ij} - k), \tag{4.5}$$

where a set of cross filters $\mathbf{w}_{ij}$ performs the unmixing and deconvolution, while a single feedforward weight, w_i performs a scaling operation (gain control).

Although he presented good results, this architecture can fail in principle. This is because each 'leading' cross weight w_{ij1} carries activation about the mixtures at the previous time step, $\mathbf{x}(t-1)$. It will thus be impossible to exactly cancel out interference at u_i from the sources, $\mathbf{x}(t)$ at the current time point. In the limiting case of white sources, this rule will fail completely, since it relies on adjacent time points being correlated. For this reason, Torkkola's architecture is extended to a full matrix of feedback filters, including 'instantaneous' weights. Such a system may be written as

$$u_i(t) = x_i(t) - \sum_{j=1}^{N} \sum_{k=0}^{M-1} w_{ijk} u_j(t - d_{ij} - k), \tag{4.6}$$

and is shown in figure 4.1. For the reason of simplicity, the figure 4.1 shows the case for two sensors and two sources. Because terms in $u_i(t)$ appear on both sides of the equation, this system can be written in vector terms

$$\mathbf{u}(t) = \mathbf{x}(t) - \mathbf{W}_0\mathbf{u}(t) - \sum_{k=1}^{M-1} \mathbf{W}_k\mathbf{u}(t-k), \tag{4.7}$$

and in order to solve it as follows

$$\mathbf{u}(t) = (\mathbf{I} + \mathbf{W}_0)^{-1}(\mathbf{x}(t) - \sum_{k=1}^{M-1} \mathbf{W}_k\mathbf{u}(t-k)). \tag{4.8}$$

In these equations, there is a feedback unmixing matrix, $\mathbf{W}_k$, for each time point of the filter, but the 'leading matrix', $\mathbf{W}_0$ has a special status in solving for $\mathbf{u}(t)$. The delay terms, d_{ij} have disappeared, partly due to awkwardnesses introduced in moving to vector notation. The delays may be considered to be simply zero-taps, and therefore can be thought of as part of the filter. However, it is often convenient to parameterize them separately, since one meter of distance in air at an 8 kHz sampling rate, corresponds to a whole 25 taps of a filter. To this end, the delays are reintroduced in a position between the 'leading matrix' and the rest of the filters, giving the following equivalent system

$$\mathbf{u}(t) = (\mathbf{I} + \mathbf{W}_0)^{-1}(\mathbf{x}(t) - \mathbf{net}(t)), \tag{4.9}$$

$$net_i(t) = \sum_{j=1}^{N} \sum_{k=1}^{M-1} w_{ijk} u_j(t - d_{ij} - k)). \tag{4.10}$$

4.3.1 Learning Rules

Learning in this architecture is performed by maximizing the joint entropy, $H(\mathbf{y}(t))$, of the random vector $\mathbf{y}(t) = g(\mathbf{u}(t))$, where g is a bounded monotonic nonlinear

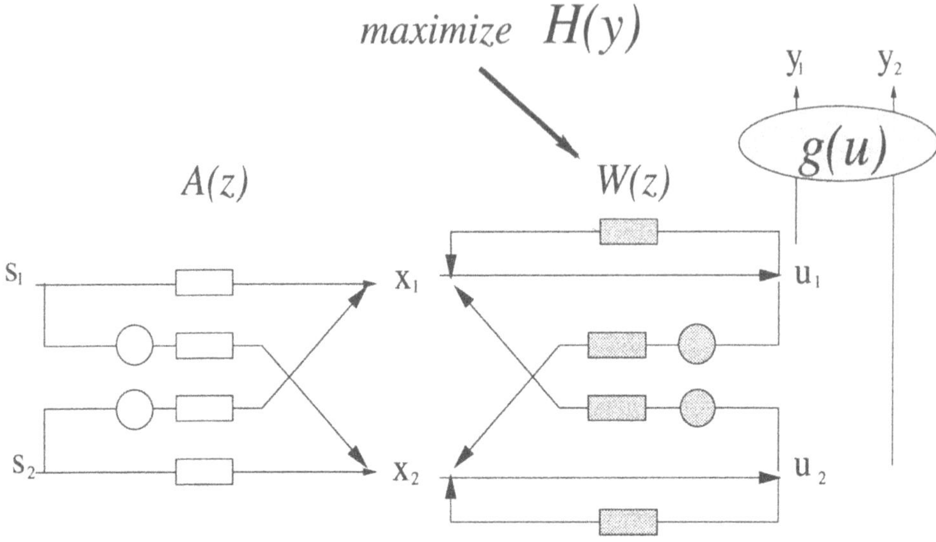

Figure 4.1. The feedback architecture of eq.4.13, which is used to separate and deconvolve signals. Each box represents a causal filter and a circle denotes a time-delay.

function (a sigmoid function). In the static feedback case of eq.4.10, when $M = 1$, the learning rule for the feedback weights $\mathbf{W}_0$ is just a coordinate transform of the rule for feedforward weights, $\hat{\mathbf{W}}$, in the equivalent architecture of $\mathbf{u}(t) = \hat{\mathbf{W}}\mathbf{x}(t)$. Since $\hat{\mathbf{W}} \equiv (\mathbf{I} + \mathbf{W}_0)^{-1}$, it follows that $\mathbf{W}_0 = \hat{\mathbf{W}}^{-1} - \mathbf{I}$, which, due to the quotient rule for matrix differentiation, differentiates as

$$\Delta\mathbf{W}_0 = -(\hat{\mathbf{W}}^{-1})\Delta\hat{\mathbf{W}}(\hat{\mathbf{W}}^{-1}). \tag{4.11}$$

Eq.4.11 can be found by a first order approximation of a small perturbation in $\mathbf{W}$ as follows

$$\begin{aligned} \hat{\mathbf{W}}\hat{\mathbf{W}}^{-1} &= \mathbf{I} \\ (\hat{\mathbf{W}} + \Delta\hat{\mathbf{W}})(\hat{\mathbf{W}}^{-1} + \Delta\hat{\mathbf{W}}^{-1}) &= \mathbf{I} \\ \hat{\mathbf{W}}\hat{\mathbf{W}}^{-1} + \hat{\mathbf{W}}\Delta\hat{\mathbf{W}}^{-1} + \Delta\hat{\mathbf{W}}\hat{\mathbf{W}}^{-1} + \Delta\hat{\mathbf{W}}\Delta\hat{\mathbf{W}}^{-1} &= \mathbf{I} \\ \hat{\mathbf{W}}\Delta\hat{\mathbf{W}}^{-1} + \Delta\hat{\mathbf{W}}\hat{\mathbf{W}}^{-1} &= \mathbf{I} \\ \Delta\mathbf{W}_0 \equiv \Delta\hat{\mathbf{W}}^{-1} &= -(\hat{\mathbf{W}}^{-1})\Delta\hat{\mathbf{W}}(\hat{\mathbf{W}}^{-1}). \end{aligned} \tag{4.12}$$

Substituting into eq.2.45 gives the natural gradient [3] rule for static feedback weights

$$\Delta\mathbf{W}_0 \propto -(\mathbf{I} + \mathbf{W}_0)(\mathbf{I} + \hat{\mathbf{y}}\mathbf{u}^T), \tag{4.13}$$

where the elements of $\hat{\mathbf{y}}$ are

$$\hat{y}_i = \frac{\partial}{\partial y_i}\frac{\partial y_i}{\partial u_i}. \tag{4.14}$$

In case of a logistic activation function

$$\hat{y}_i = 1 - 2\frac{1}{1+\exp(-u_i)}, \tag{4.15}$$

the learning rule is then suited to separate super-Gaussian sources. However, one can formulate $\hat{\mathbf{y}}$ so that it takes into account both distributions sub- and super-Gaussian sources (Lee et al., 1998b)

$$\hat{\mathbf{y}}_{ss} = -\mathbf{K}\tanh(\mathbf{u}) - \mathbf{u} \tag{4.16}$$

where k_i are elements of the N-dimensional diagonal matrix $\mathbf{K}$. The switching parameter k_i can be derived from the generic stability analysis of separating solutions as employed in chapter 2. For the reason of simplicity, this chapter considers the separation of super-Gaussian sources, hence eq.4.15, only. The extension to sub- and super-Gaussians is straightforward.

The procedure for instantaneous feedback weights may be extended to networks involving filters. For the feedforward filter architecture $\mathbf{u}(t) = \sum_{k=0}^{M-1}\hat{\mathbf{W}}_k\mathbf{x}(t-k)$, the natural gradient rule (for $k > 0$) is

$$\Delta\hat{\mathbf{W}}_k \propto (\hat{\mathbf{y}}\mathbf{u}_{t-k}^T - \mathbf{u}\mathbf{u}_{t-k}^T)\hat{\mathbf{W}}_k, \tag{4.17}$$

where, for convenience, time has become subscripted. Performing the same coordinate transforms as for $\mathbf{W}_0$ above, gives the rule

$$\Delta\mathbf{W}_k \propto -(\mathbf{I}+\mathbf{W}_k)\hat{\mathbf{y}}\mathbf{u}_{t-k}^T. \tag{4.18}$$

Finally, for the delays in eq.4.10, the derivation leads to

$$\Delta d_{ij} \propto \frac{\partial H(\mathbf{y})}{\partial d_{ij}} = -\hat{y}_i\sum_{k=1}^{M-1}\frac{\partial}{\partial t}w_{ijk}u(t-d_{ij}-k). \tag{4.19}$$

This rule is different from that in Torkkola (1996a) because it uses the collected temporal gradient information from all the taps. The algorithms of eq.4.13, eq.4.18 and eq.4.19 are the ones we use in the following simulation experiment on the architecture of eq.4.10.

4.3.2 Simulations

To verify the convergence of the learning rules in eq.4.13, eq.4.18 and eq.4.19 an IIR filter system is used as shown in figure 4.1. The super-Gaussian white noise sources were generated artificially and then mixed and delayed in the time domain as follows

$$\begin{aligned} A_{11}(n) &= 0.9 + 0.5n^{-1} + 0.3n^{-2} \\ A_{12}(n) &= -0.7n^{-5} - 0.3n^{-6} - 0.2n^{-7} \\ A_{21}(n) &= 0.5n^{-5} + 0.3n^{-6} + 0.2n^{-7} \\ A_{22}(n) &= 0.8 - 0.1n^{-1}. \end{aligned} \tag{4.20}$$

The mixing system, $\mathbf{A}(z)$, is determined by transforming the filter coefficients in the frequency domain and forming the determinant of the filter matrix. By computing the roots of the determinant, the zeros of the system A(z) can be determined. For the minimum-phase system all zeros are inside the unit circle. In the case of eq.4.20 A(z) had all zeros inside the unit circle. Hence, $\mathbf{A}(z)$ can be inverted using a stable causal IIR system since all poles of the inverting systems are inside the unit circle. In the frequency domain the weight filters have the following form

$$\begin{bmatrix} U_1(z) \\ U_2(z) \end{bmatrix} = \begin{bmatrix} W_{11}(z) & W_{12}(z) \\ W_{21}(z) & W_{22}(z) \end{bmatrix} \begin{bmatrix} X_1(z) \\ X_2(z) \end{bmatrix}. \tag{4.21}$$

This leads to the following solution for the weight filters

$$\begin{array}{ll} W_{11}(z) = \frac{1}{\Delta(z)} A_{22}(z) & W_{22}(z) = \frac{1}{\Delta(z)} A_{11}(z) \\ W_{12}(z) = -\frac{1}{\Delta(z)} A_{21}(z) & W_{21}(z) = -\frac{1}{\Delta(z)} A_{12}(z), \end{array} \tag{4.22}$$

where $\Delta(z) = W_{11}(z)W_{22}(z) - W_{12}(z)W_{21}(z))$. Figure 4.2 (a) to (d) shows the learned filters and the time-delays. Figure 4.2 (e) and (f) show the overall performance of the mixing and unmixing system. Note that the figures 4.2 (e) and (f) show the direct filter path as well as the cross filter path.

For this simulation, artificial data was used and the time-delays are easily learned simultaneously. However, for natural signals such as speech signals the delay learning rule depends on initial values of the learning algorithm. Due to the high correlation between natural signals local minima and maxima may occur in the entropy over time-delays diagram.

Simulations are performed on adaptive delays for the mixing of two speech signals. The speech signals (10 sec each) were mixed together statically by choosing M=1 with the cross delay values D_{12} and D_{21} fixed at $(25, 25)$. By varying the delays D_{12} and D_{21} the surface of the entropy over a range of ± 25 time-delays can be analyzed as shown in figure 4.3. The absolute maximum is reached at D(25,25) which corresponds to the maximum entropy at the correct learned delay values. The minima and maxima in figure 4.3 indicate the high correlation between two speech signals.

4.4 FEEDFORWARD ARCHITECTURE

The feedforward architecture is shown in figure 4.4 and can be described as

$$u_i(t) = \sum_{j=1}^{N} \sum_{k=0}^{M-1} w_{ijk} x_j(t-k). \tag{4.23}$$

In this, there are filters, $\mathbf{w}_{ij}$, which supposedly reproduce, at the u_i, the original uncorrupted source signals, s_i. This was the architecture implicitly assumed in Bell and Sejnowski (1995). The time-delays are not learned separately due to the reasons explained in the previous section. However, the time-delays can be incorporated as part of the deconvolving filter.

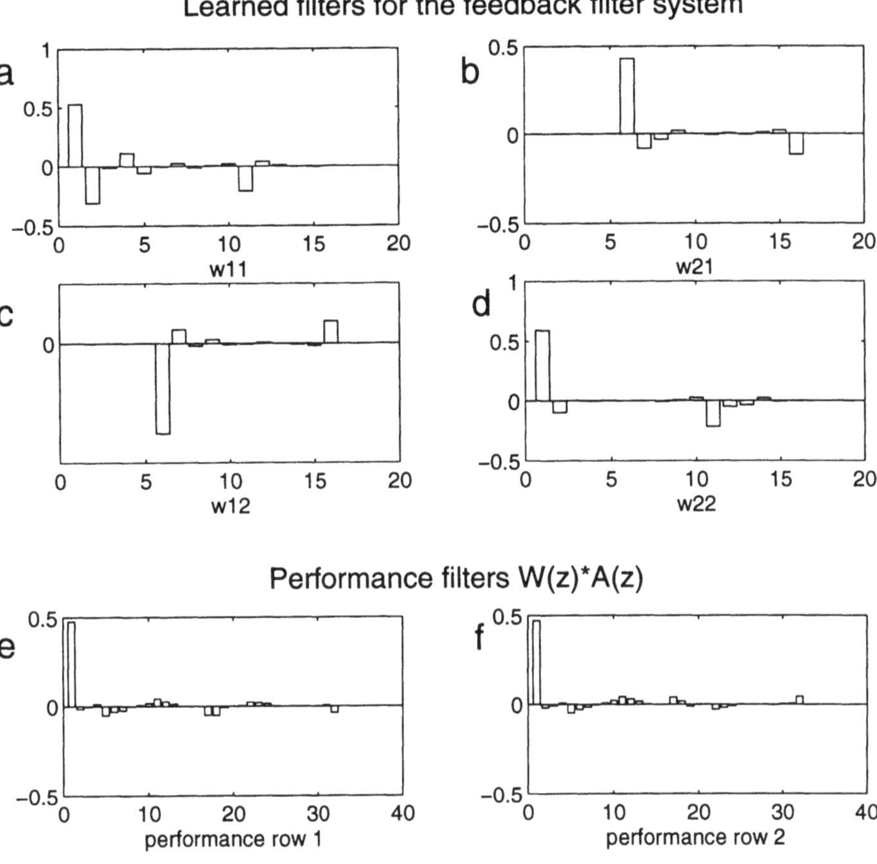

Figure 4.2. Learned Filters and time-delays $W(z)$ (a)-(d) according to eq.4.13, eq.4.18 and eq.4.19 for the full feedback system and the overall performance for both rows (e) and (f).

The advantage of the feedforward system is that it can learn a more general inverse system since it can approximate a solution for non-minimum phase mixing systems $\mathbf{A}(z)$. For example, a non-minimum phase system will occur when a microphone picks up an echo that is stronger than the direct signal. Then the increase in negative phase is directly related to the amount of temporal delay of a narrowband component at that frequency. Hence, the minimum phase lag property or the minimum group delay property of a non-minimum-phase system is not guaranteed. Since one cannot obtain prior knowledge about the mixing properties in real recordings it is more general to assume a non-minimum phase system which may have a non-causal filter system inverse. Strictly non-causal filters (dependency on an infinite number of past time-samples) cannot be implemented. However, any non-minimum or true phase system can be expressed as $H(z) = H_{min}(z)H_{AP}(z)$ where $H_{min}(z)$ is a minimum

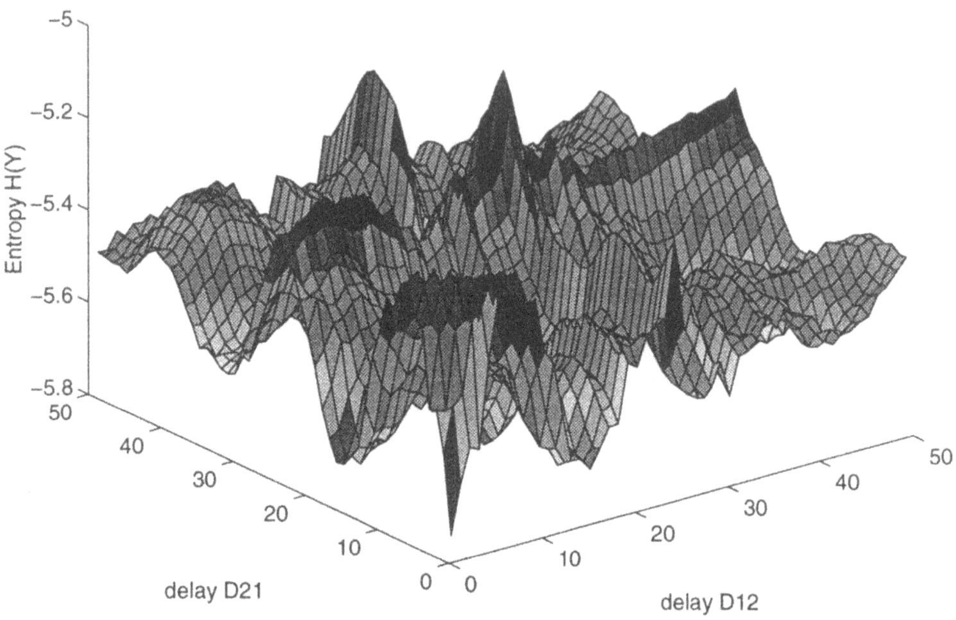

Figure 4.3. Entropy as a function of the delays D_{12} and D_{21}. Two linear mixed speech samples were used to compute the entropy surface.

phase system and $H_{AP}(z)$ is an *all-pass* system. $H_{min}(z)$ has all its poles and zeros inside the unit circle and $H_{AP}(z)$ represents a time delay with a unit frequency magnitude response. Therefore, $H_{AP}(z)$ preserves the amplitude frequency spectrum and imposes a time delay on $H(z)$ by reflecting the zeros outside the unit circle to their conjugate reciprocal location inside the unit circle. By time-delaying the inverting system up to $M/2$ taps, M being the size of the inverting filter, a $M/2$-order $H_{AP}(z)$ filter is incorporated. This is a technique to realize non-causal systems.

4.4.1 *Learning Rules*

The learning rules for the feedforward system can be derived in the same manner as for the feedback case. The mixing and unmixing system in figure 4.4 can be written in the frequency domain representation where the elements of the matrices are filters. Then, the multiplication operation replaces the convolution property. Lambert (1996) and Lambert and Nikias (1995b, 1995a) showed that FIR polynomial matrix algebra can be used as an efficient tool to elegantly solve problems for the

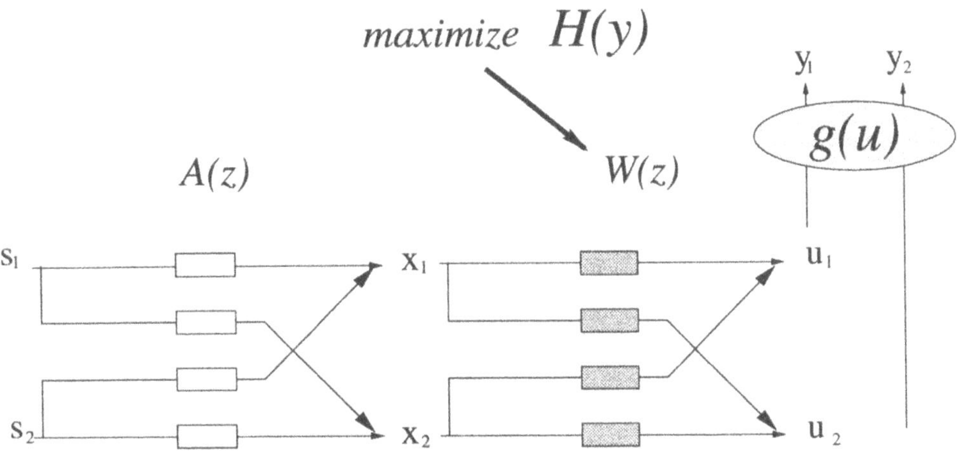

Figure 4.4. The feedforward architecture of eq.4.23, which is used to separate and deconvolve signals. Each box represents a causal or non-causal filter.

multichannel source separation. The goal of using the FIR polynomial matrix algebra is to extend the algebra of scalar matrices to the algebra of matrices of filters (time-domain) or polynomials (freq. domain). The methods for computing functions of an FIR filter, such as an inverse, involve the formation of a circulant data matrix. Due to this nature one can move to the frequency domain representation where eigencolumns of the circulant matrix are the discrete Fourier basis functions of the FFT of corresponding length. The filters now become polynomials of the Laurent series extension (z-transform) and the convolution and deconvolution of filters is reduced to multiplication and division of polynomials. For example, the inverse of a filter $w(t)$ is such a computation and can be formulated as follows

$$w(t)^{-1} = \mathtt{FFTSHIFT}(\mathtt{IFFT}(\mathtt{FFT}[000\cdots w(t)\cdots 000])^{-1})). \tag{4.24}$$

The prepending of postpending of zeros is needed to produce a good estimate of the double-sided Laurent series expansion to allow for non-causal expansions of non-minimum phase roots. The circular reordering in the time domain shifts the zeroth lag to the center of the filter (FFTSHIFT). Lambert (1996) presents a complete proof and justification of FIR polynomials. The learning algorithm for the two sources and two sensors problem can be reformulated from eq.2.45 as follows

$$\begin{aligned}\Delta \mathbf{W}(z) &= \\ & \left(\begin{bmatrix} \bar{1} & \bar{0} \\ \bar{0} & \bar{1} \end{bmatrix} + \begin{bmatrix} \mathtt{FFT}(\hat{y}_1) \\ \mathtt{FFT}(\hat{y}_2) \end{bmatrix} \begin{bmatrix} \mathtt{FFT}(u_1) & \mathtt{FFT}(u_2) \end{bmatrix}^* \right) \\ & \times \begin{bmatrix} W_{11}(z) & W_{21}(z) \\ W_{12}(z) & W_{22}(z) \end{bmatrix},\end{aligned} \tag{4.25}$$

where $\bar{1}$ and $\bar{0}$ denote vectors (of the length of the FFT operation) of ones and zeros respectively. Note that the neural processor $\hat{y}_i$ still operates in the time domain and the FFT is applied at the output and $*$ denotes the complex conjugate form. The more general form to separate sub- and super-Gaussian sources is

$$\Delta\mathbf{W}(z) = \left(\begin{bmatrix} \bar{1} & \bar{0} \\ \bar{0} & \bar{1} \end{bmatrix} + \begin{bmatrix} \mathrm{FFT}(\hat{y}_1) \\ \mathrm{FFT}(\hat{y}_2) \end{bmatrix} \begin{bmatrix} \mathrm{FFT}(u_1) & \mathrm{FFT}(u_2) \end{bmatrix}^* - \begin{bmatrix} \mathrm{FFT}(u_1) \\ \mathrm{FFT}(u_2) \end{bmatrix} \begin{bmatrix} \mathrm{FFT}(u_1) & \mathrm{FFT}(u_2) \end{bmatrix}^* \right) \times \begin{bmatrix} W_{11}(z) & W_{21}(z) \\ W_{12}(z) & W_{22}(z) \end{bmatrix}, \tag{4.26}$$

where now $\hat{y}_i = -K_i i \tanh(u_i)$. Eq.4.26 is analogous to eq.2.56 for the instantaneous case. Both equations, Eq.4.25 and eq.4.26, are of the form of the least mean squared (LMS) adaptive filters. A fast implementation of the LMS adaptive filters in the frequency domain can be achieved by employing the *overlap and save* block LMS technique (Oppenheim and Schafer, 1989), i.e. two blocks are processed simultaneously and x_k is shifted by one block after each iteration.

$$X(z) = \mathrm{FFT}[x_{(k-1)n} \cdots x_{kn-1} x_{kn} \cdots x_{kn+n-1}]. \tag{4.27}$$

For a block size of 1024 FFT-points the method is 16 times faster than the conventional LMS method (Ferrara, 1980).

4.4.2 Simulations

To verify the learning algorithm in eq.4.25 the FIR mixing filter system in figure 4.4 was used in which the sources had been mixed with a non-minimum-phase system. A slight modification of the mixing system in eq.4.20 was necessary to transform a minimum phase system to a non-minimum phase system. This was achieved by changing the filter $A_{11}(n)$ as follows

$$A_{11}(n) = 1 + 1.0n^{-1} - 0.75n^{-2}. \tag{4.28}$$

The mixing system has now a zero outside the unit circle as shown in figure 4.5. The pole-zero diagram is computed by taking the roots of the polynomial in eq.4.28. In figure 4.6 the learned filter system $\mathbf{W}(z)$ for inverting the mixing system in eq.4.20 are shown. Compared to the filters in figure 4.2 the learned filters have non-zero filter taps before the leading taps at half the filter size. The implementation of $M/2$ filter taps is a way to stabilize a resulting non-causal solution. The leading weights are chosen to be at half the filter size ($M/2$). Non-causality of the filters are clearly observed for W_{12}, W_{21} where non-zero coefficients are observed in front of the leading weights. Figure 4.6 (e) and (f) show the overall performance for the mixing and unmixing system, $\mathbf{P}(z) = \mathbf{A}(z) * \mathbf{W}(z)$.

4.5 EXPERIMENTS IN REAL ENVIRONMENTS

Several experiments in a normal office room (3 m x 4 m) and a conference room (8 m x 5.5 m) were conducted. The position of the two distant talking microphones [4]

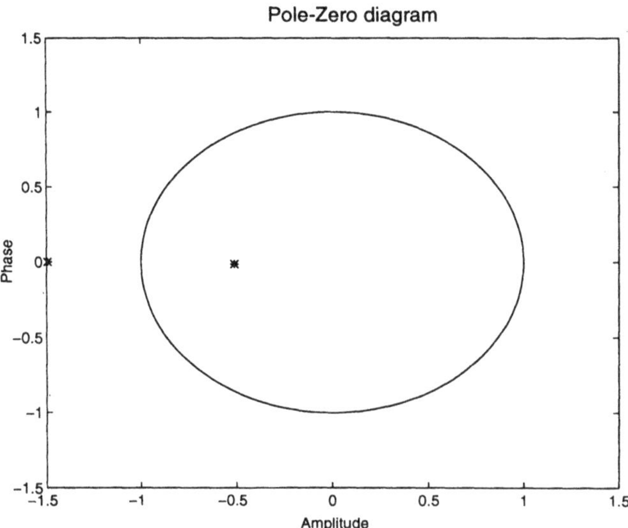

Figure 4.5. The pole-zero diagram of a non-minimum phase mixing system in eq.4.28. The filter has one zero outside the unit circle. An inverting system requires a pole outside the unit circle which leads to instability in a feedback system.

and the location of the sources had been varied for each experiment. In the first set of experiments one speaker saying the digits from one to ten while loud music was playing in the background was recorded. In this experimental setup the sources and sensors were placed in a rectangular (60 cm x 40 cm) order with 60 cm distance between the sources and the sensors.

Figure 4.7(a) and (b) shows the recorded signals where the speech signal had been heavily corrupted by a music signal. The learning algorithm in eq.4.25 was applied to a recording of 7 sec sampled at 16 kHz (120000 time points). The learning rate was fixed at 0.0001; a momentum term and a block size of 256 data points was used. The unmixed signals were obtained using 128 taps FIR filters which cover a delay of 3.2 ms corresponding roughly to 1 m. The algorithm converged in 30 passes through the data. The separated signals are shown in figure 4.7 (c) and (d). A listening test showed a clean speech separation. The learned filters are shown in figure 4.8. Better separation quality were obtained with longer filters. In figure 4.9 the learned filters are shown when a large block size of 16000 data points and 1024 taps for each filter was used. This filter can now cover a delay of about 32 ms (about 10 m). A larger block size is beneficial to minimize the influence of variances in the speech signal.

In each filter, the leading tap is followed by a strong negative tap which indicates that the infomax algorithm tries to decorrelate adjacent time points. This effect is called whitening and it increases the energy in the higher frequency spectrum and

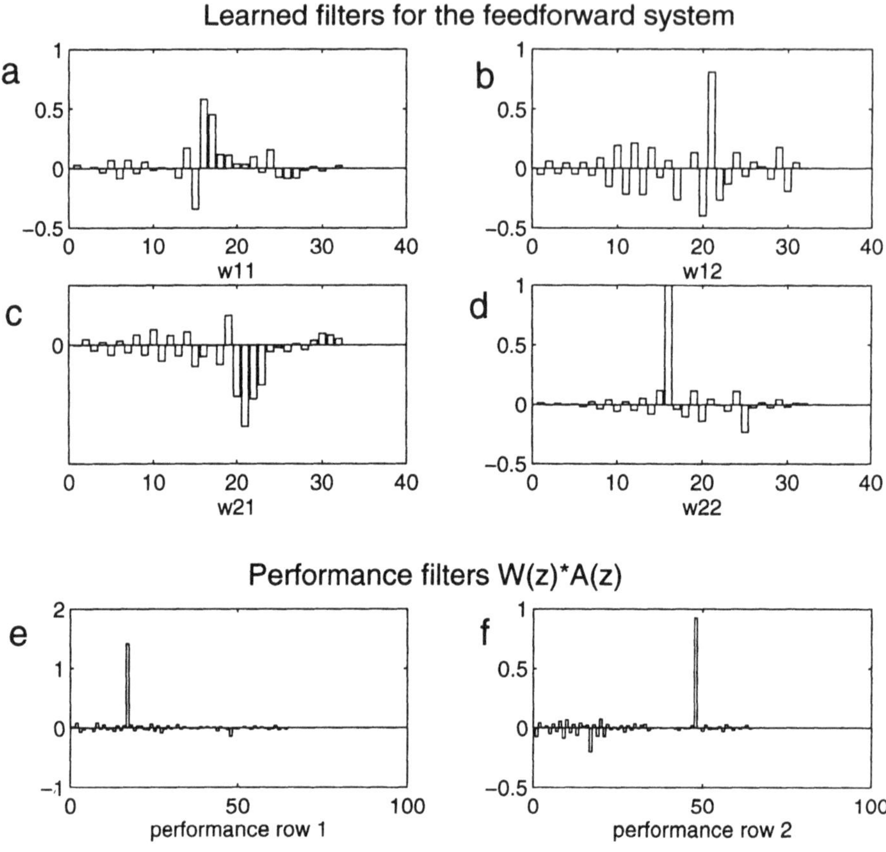

Figure 4.6. Learned Filters $\mathbf{W}(z)$ (a)-(d) for the feedforward system using FIR polynomial matrix algebra in eq.4.25 and the overall performance for both rows (e) and (f). (e) shows $[P_{11}(z), P_{21}(z)]$ and (f) shows $[P_{12}(z), P_{22}(z)]$.

reduces the energy in the lower frequency band. Whitened speech signals sound sharper than their original. This effect can be compensated by postprocessing the unmixed signals with a dewhitening filter.

Another set of experiments were performed with two people speaking simultaneously. Figure 4.10 (a) and (b) show the signals recorded with the same setup but with another person saying the digits one to ten in Spanish *(uno dos* $\cdots$ *diez)* instead of the music source. The separated signals are shown in figure 4.10 (c) and (d). A listening test shows an almost clean speech separation [5]. An objective measure for the quality of separation is difficult to obtain since the original signals are not available. Human perception usually varies and a more objective measure are results from an automatic speech recognition system.

Figure 4.7. Cartoon of the cocktail party problem. People are talking while music is playing in the background. In an experiment in a normal office room (3m x 4m) the voice of one person and the background music was recorded with two microphones. The sources and sensors were placed in a rectangular (60 cm x 40 cm) order with 60 cm distance between the source and the microphone.

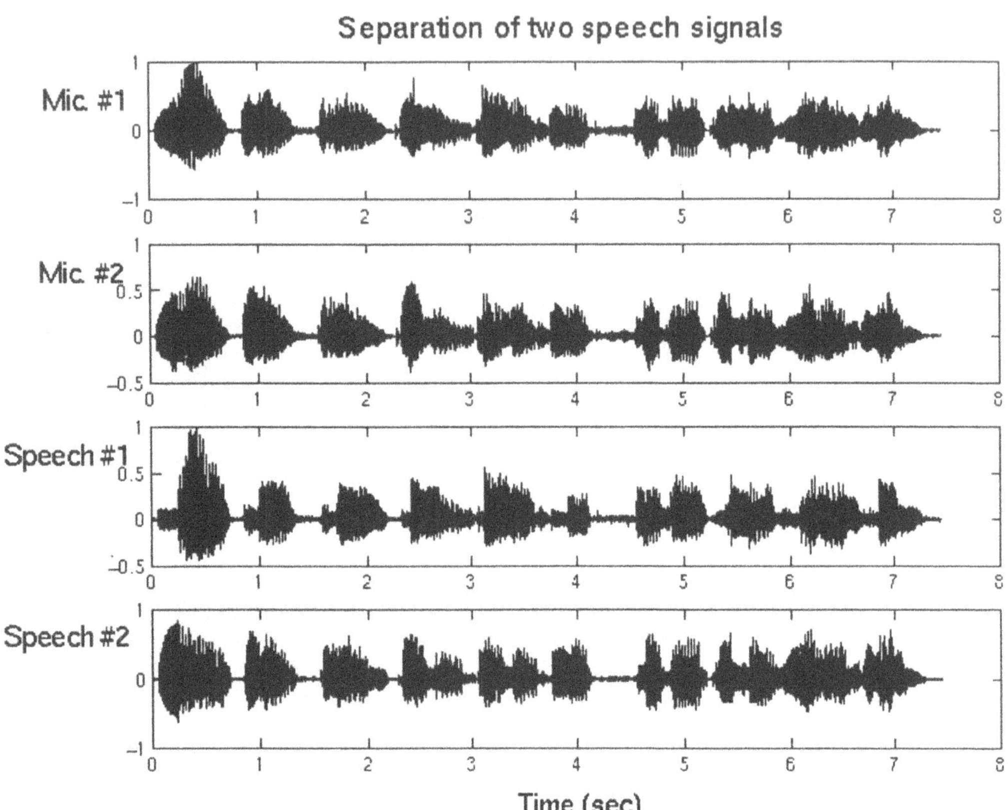

Figure 4.8. Microphone outputs of one voice with background music recorded in a normal office room. The recordings are (a) Microphone 1 (b) Microphone 2 and the separated signals are (c) speech and (d) music.

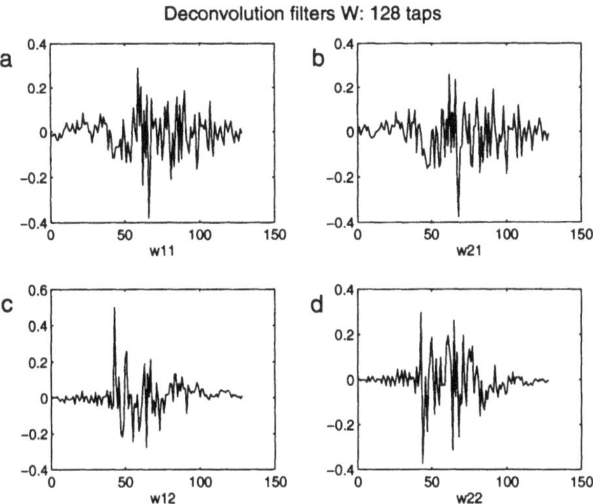

Figure 4.9. FIR 128-tap filters that unmixed and deconvolved the speech and music signals. Leading weights of the channel filters W_{11} and W_{22} are at 64-taps. Cross-channel filters are W_{21} and W_{12}.

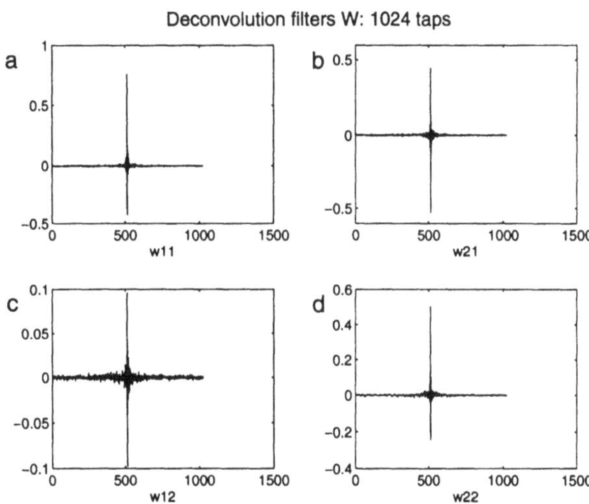

Figure 4.10. FIR 1024-tap filters that unmixed and deconvolved the speech and music signals. Leading weights of the channel filters W_{11} and W_{22} are at 512-taps. Cross-channel filters are W_{21} and W_{12}.

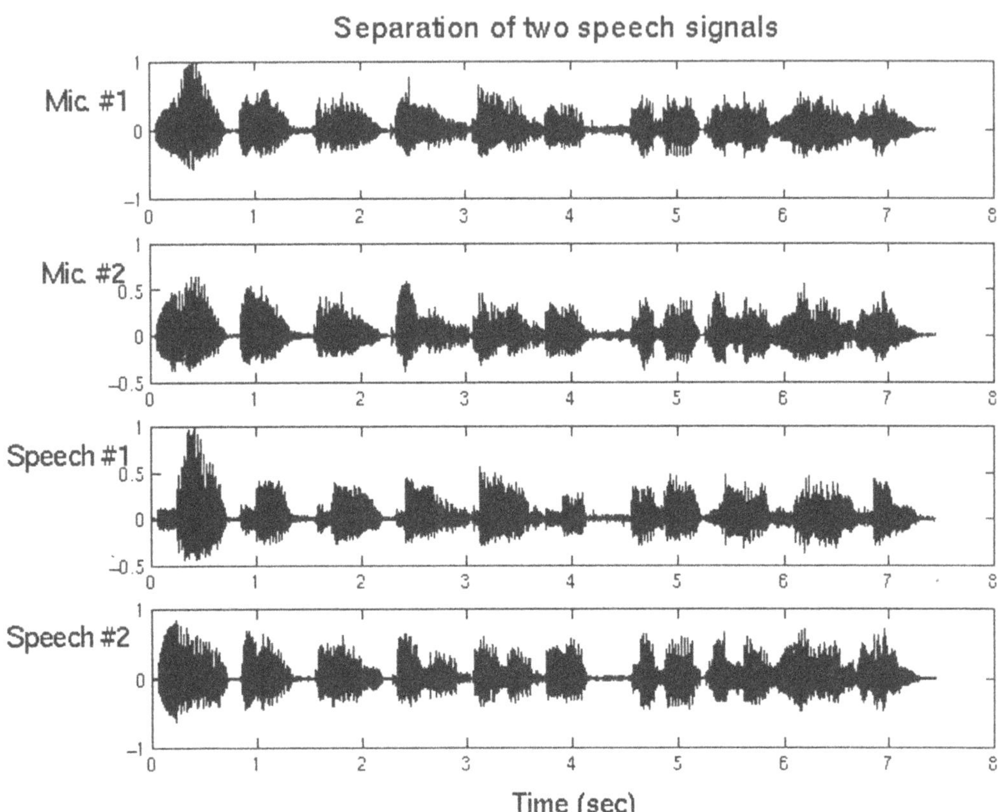

Figure 4.11. Microphone outputs of two speakers recorded in a normal office room (a) microphone 1 and (b) microphone 2. The separated speakers are shown in (c) Spanish digits *uno dos ... diez* and (d) English digits *one two ... ten.*

Table 4.1. Speech recognition results [recognition rate in %].

Recog. rate	No. of words	mixtures	separation
Speech-Music	100	14 %	64 %
Speech-Speech	100	42 %	61 %
TOTAL	200	28 %	62.5 %

4.5.1 Speech Recognition Results

A prospective application of the multichannel blind separation algorithms are spontaneous speech recognition [6] tasks. The best speech recognizer may fail completely in the presence of background music or competing speakers as in the teleconferencing problem. An automatic speech recognition system was used to obtain speech recognition results on the separated speech signals. The recognizer was trained on the *Wall Street Journal task* [7]. The recognition results are listed in table 4.5.1. High recognition errors rates were obtained when the microphone outputs were applied to the recognition engine. Significantly row recognition error rates were obtained once the separated signals were presented to the recognizer. The results can be further improved by postprocessing the separated signals, e.g., zeroing out the noisy part with a low signal-power detector and by using a speech recognizer trained on digits.

4.6 BUSSGANG ALGORITHMS

The goal of this section is to present an alternative methods to the infomax algorithm. Another way of deriving the infomax learning rules from the Bussgang property was described in chapter 2. This section briefly summarizes how the infomax learning rules for multichannel source separation problems can be seen from the Bussgang property. Lambert (1996) derived three classes of direct Bussgang cost functions. The Bussgang property in the frequency domain is as follows

$$E\{\mathbf{U}(\mathbf{z})\mathbf{U}(\mathbf{z})^H\} = E\{\hat{\mathbf{Y}}(\mathbf{z})\mathbf{U}(\mathbf{z})^H\}, \tag{4.29}$$

where $\mathbf{U}(\mathbf{z})$ are the estimated source in the frequency representation, $\hat{\mathbf{Y}}(\mathbf{z})$ are the frequency representations of $\hat{\mathbf{y}}(\mathbf{u})$ in eq.4.15 and the superscript H denotes the conjugate transpose. The Bussgang form in eq.4.29 can be rewritten as

$$E\{\mathbf{U}(\mathbf{z})\mathbf{X}(\mathbf{z})^T\} = E\{\hat{\mathbf{Y}}(\mathbf{z})\mathbf{X}(\mathbf{z})^T\}, \tag{4.30}$$

which gives the first direct Bussgang algorithm cost function. The update for the weight filters is (Lambert, 1996)

$$\mathbf{W}(z)_{n+1} = \mathbf{W}(z)_n + \epsilon(\mathbf{U}(\mathbf{z}) - \hat{\mathbf{Y}}(\mathbf{z}))\mathbf{X}(\mathbf{z})^T, \tag{4.31}$$

and its form is similar to the form of the blind least mean squared error method. Eq.4.29 can be rewritten in the second form

$$\mathbf{W}(z)\mathbf{A}(z)E\{\mathbf{S}(\mathbf{z})\mathbf{S}(\mathbf{z})^T\}\mathbf{A}(z)^H = E\{\hat{\mathbf{Y}}(\mathbf{z})\mathbf{X}(\mathbf{z})^T\}. \tag{4.32}$$

Since at convergence the terms are $(\mathbf{S(z)S(z)}^T = \mathbf{I}(z))$ and $(\mathbf{W}(z)\mathbf{A}(z) = \mathbf{I}(z))$ where $\mathbf{I}(z)$ is an identity matrix of filters, second form for the Bussgang algorithm cost function can be written as

$$\mathbf{W}(z)_{n+1} = \mathbf{W}(z)_n + \epsilon((\mathbf{W(z)^H})^{-1} - \hat{\mathbf{Y}}(\mathbf{z}))\mathbf{X}(\mathbf{z})^T, \tag{4.33}$$

which is of the same form of the original infomax algorithm (Bell and Sejnowski, 1995). A third form of the direct Bussgang algorithm can be formulated as follows (Lambert, 1996)

$$E\{\mathbf{U(z)U(z)}^T\}\mathbf{W}(z) = E\{\hat{\mathbf{Y}}(\mathbf{z})\mathbf{U}(\mathbf{z})^T\}\mathbf{W}(z), \tag{4.34}$$

where $E\{\mathbf{U(z)U(z)}^T\}$ converges to $\mathbf{I}(z)$ for the correct weight matrix $\mathbf{W}(z) = \mathbf{A}(z)^{-1}$. The third form leads to a new weight update (Lambert, 1996)

$$\mathbf{W}(z)_{n+1} = \mathbf{W}(z)_n + \epsilon(\mathbf{I}(z) - \hat{\mathbf{Y}}(\mathbf{z})\mathbf{U}(\mathbf{z})^T)\mathbf{W}(z), \tag{4.35}$$

which has exactly the form of the original infomax algorithm supplied with the natural or relative gradient. The three Bussgang algorithms are valid derivations from the Bussgang property. However, the reasoning for the Bussgang algorithm becomes clear when relating it to the infomax principle. A justification why the third form converges faster than the other forms cannot be explained by this derivation.

4.7 TIME-DELAYED DECORRELATION METHODS

There are several other approaches to the multichannel deconvolution problem. One interesting approach is the time-delayed decorrelation method which gives similar separation performance under certain conditions.

Although chapter 2 suggested that decorrelation-based methods fail to separate sources, there exist methods to separate sources using higher-order order information in different time-lags. The basic ideas is to extend the the decorrelation-based method to simultaneously time-delayed decorrelation (TDD). The goal of TDD is now to diagonalize the covariance matrix $\mathbf{C}_0 = \langle \mathbf{x}(t)\mathbf{x}(t)^T \rangle$ for $\tau = 0$ (no time-delay) and at the same time to diagonalize the covariance matrix for a given delay $\mathbf{C}_\tau = \langle \mathbf{x}(t)\mathbf{x}(t-\tau)^T \rangle$. The covariance matrix can be decomposed into a matrix $\mathbf{A}$ and a diagonal matrix Λ as follows

$$\mathbf{C}_0 = \mathbf{A}\Lambda_0\mathbf{A}^T \tag{4.36}$$

$$\mathbf{C}_\tau = \mathbf{A}\Lambda_\tau\mathbf{A}^T \tag{4.37}$$

This leads to an eigenvalue problem as described in Molgedey and Schuster (1994); Belouchrani et al. (1997)

$$(\mathbf{C}_0\mathbf{C}_\tau^{-1})\mathbf{A} = \mathbf{A}(\Lambda_0\Lambda_\tau^{-1}), \tag{4.38}$$

where the elements of Λ are the eigenvalues of the corresponding covariance matrix. The TDD algorithm can be extended to a matrix of filters (Ehlers and Schuster, 1997;

Murata et al., 1998; Lee et al., 1998e). The main extension consists of transforming the signals $x_i(t)$ into the frequency domain $X_i(z)$ and hence creating a spectrogram. A correlation matrix can be computed as in eq.4.38 for each frequency bin. The inverse of $\mathbf{A}(z)$ multiplied with the spectrogram results in the frequency domain decorrelated signals which can be reconstructed using an IFFT and overlap and zero-padding technique. The unmixing filters in the time-domain are obtained by IFFT of $\mathbf{A}(z)$. There are two optimizing steps improving the separation performance: (a) setting the direct filters to identity and therefore avoiding the whitening problem (b) optimizing a decorrelation-based cost function (Ehlers and Schuster, 1997) (c) optimizing τ as a function of decorrelation cost function. Point (c) is crucial to achieve good separation results and therefore TDD requires a secondary optimization step. The main advantage of the TDD algorithm is the computational efficiency in computing the cross-filters since no adaptation is necessary. An online-version of this algorithm could be implemented in a block mode in which successive blocks of data points (e.g. 128, 256) are processed.

A simple TDD algorithm (Molgedey and Schuster, 1994) has been shown to be highly effective under the minimum-phase constraint. The TDD algorithm can in some circumstances achieve the same separation quality much faster which is important for online implementations. The TDD formulation was introduced also earlier by other researchers, e.g. Tong et al. (1991) and Belouchrani et al. (1993). Feder et al. (1993) proposed a method for multichannel signal separation by decorrelating the outputs. Chan et al. (1996) used a decorrelation algorithm based on a constant power and a constant diagonalization constraint. Many other methods have been proposed that decorrelate the output signals (Choi and Cichocki, 1997; Douglas and Cichocki, 1997). Extensions to fourth-order decorrelation techniques were proposed by Yellin and Weinstein (1996); Nguyen-Thi and Jutten (1995) and Comon (1996).

For the multichannel blind deconvolution problem good separation results were shown. It is assumed that the results were obtained assuming a minimum-phase mixing system. In this case, the transfer function of the mixing system has all of its poles and zeros inside the unit circle in the z-plane. The feedback and feedforward unmixing system will be stable (Haykin, 1991). For a minimum-phase system, the information about the signals is preserved in the second-order statistics (Oppenheim and Schafer, 1989). Then, power spectral estimation may give a sufficient answer (Bellini, 1994). This may explain the success of simple decorrelation-based method for blind deconvolution. The minimum-phase assumption in room recordings may hold for situations where the microphones are placed close to the sources and in environments with little echos (e.g. an anechoic chamber). For most real world experiments, the general assumption that the mixing system is non-minimum-phase may be more appropriate.

4.7.1 Experimental Results with TDD

Figure 4.7.1 shows an example of a recording in a room obtained by Yellin and Weinstein (1996). Here, a music signal and a voice signal that was played by an audio system and the signals were recorded with two microphones located close to

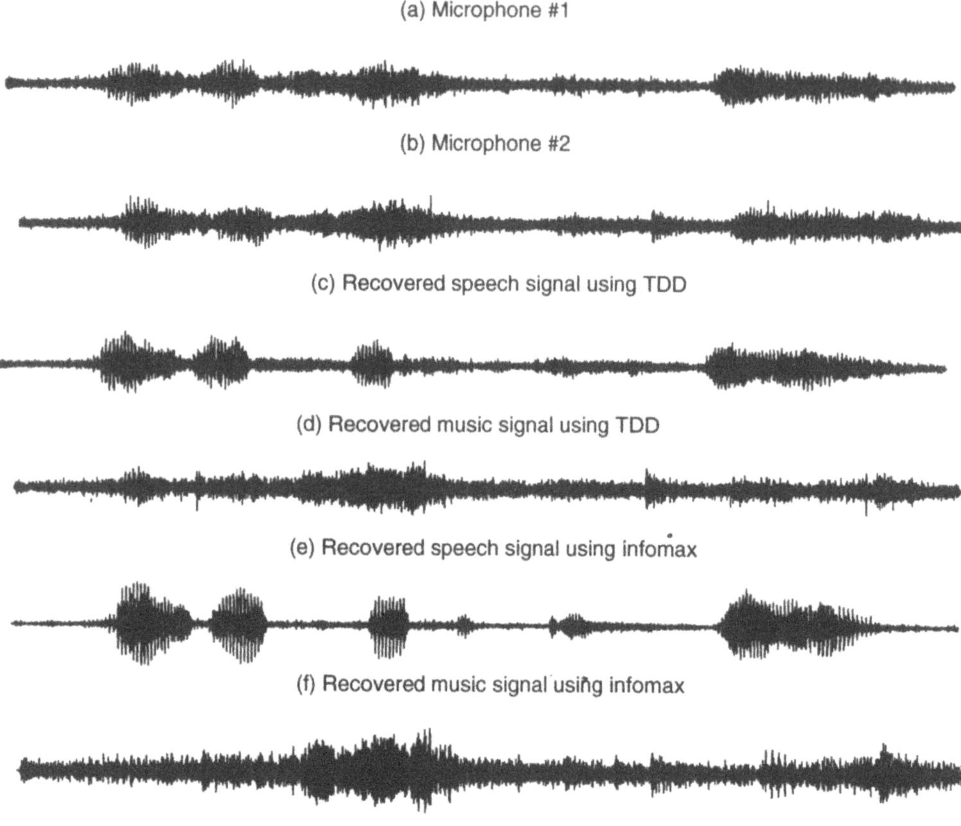

Figure 4.12. Room recordings from Yellin and Weinstein (1996): (a) microphone 1, (b) microphone 2. The separated signals using the TDD algorithm are shown for speech and music in (c) and (d). Slightly better results were obtained with eq.4.25 as shown in (e) and (f).

the sources (60 cm). Figure 4.7.1(a) and (b) show the recorded signals. Two cross filters with 128 taps each were computed using the TDD algorithm. The unmixed signals were obtained after 10 seconds on a Sparc10 workstation using MATLAB. Figure 4.7.1 (c) shows the recovered speech signal and figure 4.7.1(d) shows the music signal using the TDD algorithm by Molgedey and Schuster (1994). For the same recording the learning rule in eq.4.25 was used giving slightly better separating results shown in figure 4.7.1 (e) and (f) with the same set of parameters. Its convergence however, was slow (about 5 min with annealing the learning rate). The infomax results are very similar to the results obtained by Yellin and Weinstein (1996) using a fourth-order cumulant-based method. Unfortunately, the signal to noise ratio is not measurable due to the unavailability of the original speech and music signals. The use of the TDD algorithm as a preprocessing step for infomax approximately doubled the convergence speed. In many experiments such as the recordings by

Ehlers and Schuster (1997) and Lee et al. (1997b) the TDD algorithm by itself gave results similar in quality to infomax.

4.7.2 Discussions on TDD

Separation results of room recordings using the TDD algorithm and the infomax ICA algorithm were presented. While in general infomax achieved better separation results than the TDD algorithm, the convergence speed was slow. The TDD algorithm, however, may allow for online implementations for real-time applications such as speech recognition and may be used as a preprocessing step for infomax to speed up convergence. Additional improvements can be made in optimizing the TDD algorithm and its combination with infomax.

An extension of infomax to timed-delayed infomax was proposed by Attias and Schreiner (1998) called the dynamic component analysis (DCA). The main difference to Bell and Sejnowski's infomax principle is to ensure independent components at several different time lags. The consequence is a spatio-temporal redundancy reduction that can be used to find instantaneous unmixing matrices and filters.

4.8 SPECTROGRAM ICA

This section presents an alternative solution to multichannel blind deconvolution and its implementation. This method is similar to the dynamic component analysis by Attias and Schreiner (1998). A brief summary of the method is shown in figure 4.13. The main idea is to perform ICA on the spectrogram of the observations.

The implementation of the algorithm is demonstrated on a simple example (figure 4.13) and the steps are as follows

1. Generate spectrogram for the signals $x_1(t)$ and $x_2(t)$ using the Fourier transformation and the overlap-save technique.

2. Perform ICA on the spectrogram for each frequency bin. Since the values of the spectrogram are complex. There are two ways to solve the complex ICA problem. An extension of the infomax algorithm for complex values is presented in Smaragdis (1997). The idea is to use a complex activation function instead of the usual nonlinearity. Another method for ICA on complex valued data is the JADE algorithm (Cardoso and Soloumiac, 1993; Cardoso, 1998a).

3. The JADE algorithm uses the Jacobi rotation. This rotation is an optimization procedure that finds an optimal rotation between orthogonal matrices so that the resulting matrix is the closest orthogonal compromise between all cumulant solutions for one frequency bin.

4. The unmixed signals $\mathbf{u}_i(z)$ can now be found for each frequency bin.

5. Since the signals may be permuted and scaled. They are first rescaled and unpermuted as suggested by Murata et al. (1998). Rescaling factors are found by projecting the signals back to the observations x_i and by determining the contributions from each u_i. Permutation is corrected by finding the correlations

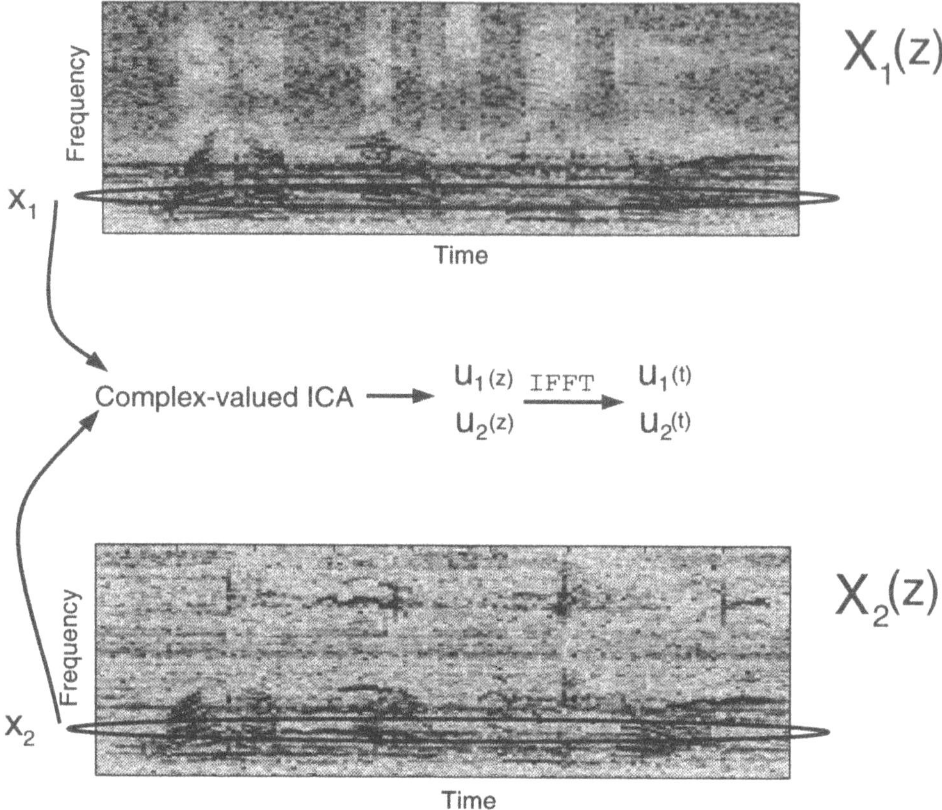

Figure 4.13. An alternative multichannel deconvolution method. A complex-valued ICA algorithm is applied to each frequency bin of the observed data. The resulting unmixed blocks are rescaled and reordered and transformed back to the time domain.

between adjacent blocks of unmixed signals. The signals u_i are reordered so that a maximum correlation coefficient is achieved for each channel.

6. The rescaled and reordered signals $\mathbf{u}(z)$ are transformed back into the time domain by using the inverse Fourier transformation and the overlap-save technique.

4.9 CONCLUSIONS

Methods were presented for the blind separation of time-delayed and convolved sources: the multichannel blind deconvolution problem. The feedback architecture gives a compact representation with respect to the number of parameters and learning rules were presented for a full feedback filter system including time-delays. Unfortunately, recordings in real environments exhibit non-minimum-phase characteristics which require a feedforward architecture for the unmixing system. The learning algorithm is efficiently updated using the polynomial filter algebra in the frequency domain. The significance of this method is shown by separating voices recorded in a normal room. This method may be used as a preprocessing step in speech recognition systems to increase the recognition rate. Other approaches are discussed and compared to the infomax approach. The TDD algorithm may allow for online implementations for real-time applications such as speech recognition and may be used as a preprocessing step for infomax to speed up convergence.

4.9.1 Future Research

The multichannel blind source separation problem is subject to further investigation due to its complexity and its problems in real world applications. Many potential applications may benefit from the new methods. In particular, communication systems such as quadrature phase shift keying (QPSK) coding schemes and code division multiple access (CDMA) with robustness to fading and noise, speech recognition systems and cross-talk elimination in telephone channels.

To this end, the following inherent problems need to be tackled for further steps towards applications:

- **Nonstationarity of Sources**

  An important observation was that slight movement of speaking people had a severe effect on the separation quality. Since the transfer function for the filter system changed over time the learned filters represented the averaged unmixing system which may results in poor separation quality depending on the degree of movements.

- **Online filter estimation**

  The number of filters increases quadratically with the number of sensors. Each filter may have 200 or more taps which signifies that the iterative estimation of all parameters ($\propto 200N^2$) is computationally expensive. Decorrelation-based algorithms converge much faster and may give good results for minimum-phase systems. Analog VLSI circuits may be developed to incorporate the multichannel deconvolution algorithm as an online preprocessing method.

- **Single-channel blind equalization**

  This problem is related to the underdetermined problem in ICA. Some limited approaches have been proposed by Nelson and Wan (1997a, 1997b) where they used a Kalman filtering approach for single channel speech enhancement. However this approach needs to be extended to filters. A multiple input and single output system (beamforming) for source separation was proposed by Li and Sejnowski (1995).The incorporation of temporal structure is key to solving the single channel source separation problem. An approach by Lewicki and Sejnowski (1998a) indicate first steps towards solving this inherently difficult problem.

- **Nonlinear Filtering**

Nonlinear phenomena may occur in the mixing process due to nonlinear characteristics of the microphones. Lee et al. (1997c) propose a method to deal with nonlinear transfer functions after a linear mixing (see next chapter). Other recently proposed nonlinear ICA approaches have not been applied to the convolved or time-delayed sources.

- **Frequency-based blind source separation**

  The polynomial filter algebra is an efficient computational tool in the frequency domain. Algorithms may be of interest that work entirely in the frequency domain. Smaragdis (1997) uses a method based on infomax in the frequency domain. The justification that infomax will work in other domains such as the Fourier domain and the wavelet domain has not been elucidated.

- **Auditory Scene Analysis**

  Auditory scene analysis is related to perception of auditory stimuli and is concerned with the questions of deciding how many sound sources there are, what are the characteristics of each source, and where each source is located. Those questions are fundamental issues in ICA and a link to the auditory scene analysis may suggest new techniques that might lead to a further understanding of the human auditory system and optimization of current techniques for finding a satisfactory solution to the cocktail-party-problem.

Notes

1. Its name is due to statistics of the deconvolved signal which are approximately Bussgang (Bellini, 1994). The Bussgang statistic refers to Bussgang at Bell Labs who found that the autocorrelation and the correlation between the signal and its nonlinearly transformed signal exhibit the same characteristics.

2. A model that can measure the power spectrum with a pole-zero transfer function

3. The natural gradient is also valid in the multichannel case (Amari et al., 1997b).

4. i.e. microphones that can record signal sources witch may be far away from the microphone.

5. These audio-files are available in `http://www.cnl.salk.edu/~tewon/`.

6. Fundamentals in speech recognition are presented by Rabiner and Juang (1993) and Deller et al. (1993).

7. The speech recognizer is trained with speech signals obtained from various people reading parts of the Wall Street Journal.

5 ICA USING OVERCOMPLETE REPRESENTATIONS

Big Brother is watching you.
George Orwell ("1984")

In this chapter empirical results are shown for the blind source separation of more sources than mixtures using a framework proposed for learning overcomplete representations recently developed by Lewicki and Sejnowski (1998b).

One of the major drawback of ICA but also one distinct feature of ICA is that the standard formulation of ICA requires at least as many sensors as sources. Lewicki and Sejnowski (1998b) have proposed a generalized ICA method for learning overcomplete representations of the data that allows for more basis vectors than dimensions in the input. This technique assumes a linear mixing model with additive noise and involves two steps: (1) learning an overcomplete basis for the observed data and (2) inferring sources given a sparse prior on the coefficients. The goal of this method is illustrated in figure 5.1. In a two-dimensional data space, the observations $\mathbf{x}$ in figure 5.1(a,b) were generated by a linear mixture of 2 independent random sparse sources. In this space, figure 5.1(a) shows orthogonal basis vectors (principle component analysis, PCA) and figure 5.1(b) shows independent basis vectors. If the 2-dimensional observed data are generated by 3 sparse sources as shown in figure 5.1(c, d) the complete ICA representation (c) cannot model the data adequately but the overcomplete ICA representation (d) finds 3 basis vectors that fit the underlying distribution of the data.

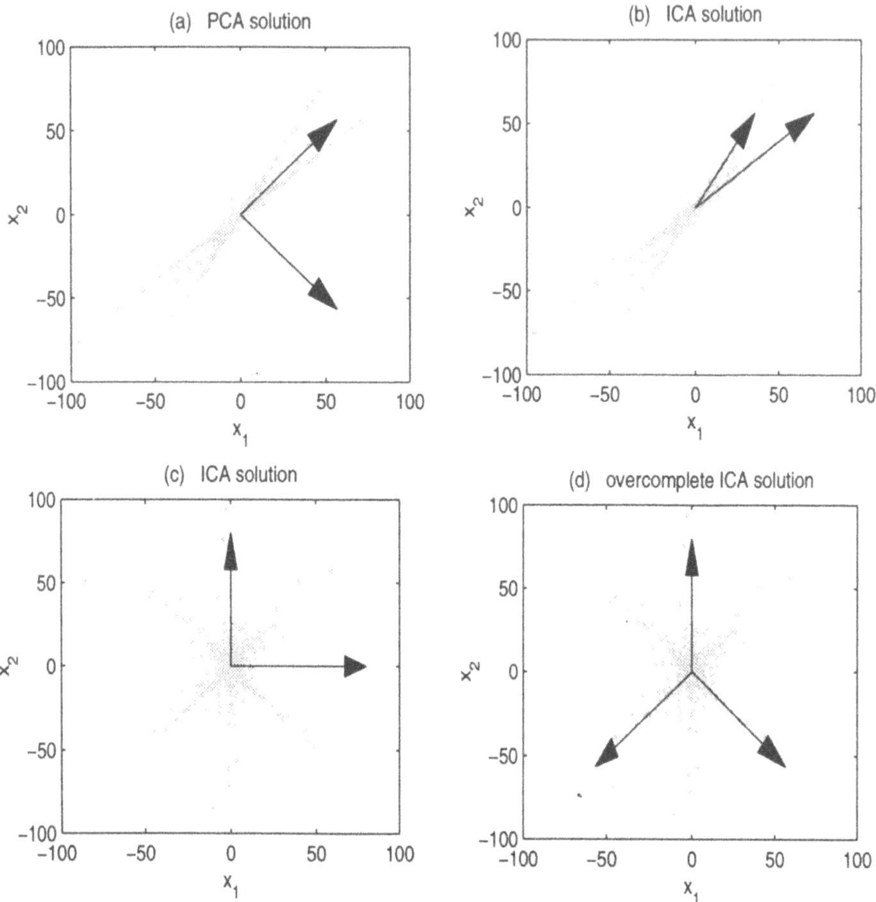

Figure 5.1. Illustration of basis vectors in a two-dimensional data space with two 2 sparse sources (top) or three sparse sources (bottom), (a) PCA finds orthogonal basis vectors and (b) ICA representation finds independent basis vectors. (c) ICA cannot model the data distribution adequately with three sources but (d) the overcomplete ICA representation finds 3 basis vectors that match the underlying data distribution (see Lewicki and Sejnowski).

In this chapter, the learning rules for overcomplete ICA are briefly summarized in section 5.1, as derived by Lewicki and Sejnowski (1998c). In section 5.2, simulation results are presented for speech signals and music signals. The discussion in section 5.3 covers related work and future research issues.

5.1 LEARNING OVERCOMPLETE REPRESENTATIONS

The observed M-dimensional data $\mathbf{x} = [x_1, \cdots, x_M]^T$ may be modeled as a linear overcomplete mixing matrix, $\mathbf{A}$, $(M \times N)$ [1] with additive noise.

$$\mathbf{x} = \mathbf{A}\mathbf{s} + \mathbf{n}, \tag{5.1}$$

where $\mathbf{s} = [s_1, \cdots, s_N]^T$ are the sources and $\mathbf{n}$ is assumed to be a white Gaussian noise with variance σ^2 so that

$$\log P(\mathbf{x}|\mathbf{A}, \mathbf{s}) \propto -\frac{1}{2\sigma^2}(\mathbf{x} - \mathbf{A}\mathbf{s}). \tag{5.2}$$

It is also assumed that the sources s_i are mutually independent, so that the joint probability distribution has the form $P(\mathbf{s}) = \prod_{i=1}^{M} P(s_i)$, and each source s_i has a sparse distribution, such as the Laplacian density $P(s_i) \propto \exp(-\alpha|s_i|)$.

Given the above model and assumptions, the goal is to infer both the basis vectors $\mathbf{A}$ and the sources $\mathbf{s}$ given the mixtures $\mathbf{x}$.

5.1.1 *Inferring the sources* $\mathbf{s}$

Due to the additive noise and the rectangular mixing matrix $\mathbf{A}$, the solution for $\mathbf{s}$ cannot be found by the pseudo-inverse $\mathbf{s} = \mathbf{A}^+\mathbf{x}$. A probabilistic approach to estimating the sources is based on finding the maximum a posteriori value of $\mathbf{s}$:

$$\begin{aligned} \hat{\mathbf{s}} &= \max_{\mathbf{s}} P(\mathbf{s}|\mathbf{x}, \mathbf{A}) \\ &= \max_{\mathbf{s}} P(\mathbf{x}|\mathbf{A}, \mathbf{s}) P(\mathbf{s}). \end{aligned} \tag{5.3}$$

Given basis vectors $\mathbf{A}$, and observation $\mathbf{x}$ eq.5.3 can be optimized by gradient ascent on the log posterior distribution (Lewicki and Sejnowski, 1998b, 1998c).

5.1.2 *Learning the basis vectors* $\mathbf{A}$

The objective for learning the basis vectors, $\mathbf{A}$, is to maximize the probability of the data which requires marginalizing over all possible sources

$$P(\mathbf{x}|\mathbf{A}) = \int P(\mathbf{x}|\mathbf{A}, \mathbf{s}) P(\mathbf{s}) d\mathbf{s}. \tag{5.4}$$

For general overcomplete bases, this integral is intractable. For the special case of zero noise and $\mathbf{A}$ invertible (a complete basis), the integral in eq.5.4 is solvable and leads to the standard ICA learning algorithm (Bell and Sejnowski, 1995; Cardoso, 1998b; Lee et al., 1998a). Lewicki and Sejnowski (1998c) approximated eq.5.4 by fitting a multivariate Gaussian around $\hat{\mathbf{s}}$. The basis vectors were learned by performing gradient ascent on the approximation to $\log P(\mathbf{x}|\mathbf{A})$:

$$\Delta\mathbf{A} \propto \mathbf{A}\mathbf{A}^T \frac{\partial}{\partial \mathbf{A}} \log P(\mathbf{x}|\mathbf{A}) \approx -\mathbf{A}(\phi(\hat{\mathbf{s}})\hat{\mathbf{s}}^T + \mathbf{I}), \tag{5.5}$$

where $\phi(\hat{s}_i) = \partial \log P(\hat{s}_i)/\partial \hat{s}_i$ is called the score function, and $\mathbf{I}$ is the identity matrix. The prefactor $\mathbf{A}\mathbf{A}^T$ produces the natural gradient extension (Amari, 1997a, 1998) which speeds convergence. Note that $\mathbf{A}$ in eq.5.5 is not restricted to be a square matrix. The derivation is described in Lewicki and Sejnowski (1998c).

5.2 EXPERIMENTAL RESULTS

5.2.1 *Blind Separation of Speech Signals*

Speech signals with silent time segments are sparsely distributed and will be approximated by a Laplacian model. Three speech signals from the same speaker, sampled at 8 kHz with 8 bits per sample, were taken from the TIMIT database and are shown in figure 5.2 (top). We mixed the three speech signals into two mixtures as follows

$$\begin{bmatrix} x_1(t) \\ x_2(t) \end{bmatrix} = \begin{bmatrix} 1 & 1/\sqrt{2} & 1/\sqrt{2} \\ 0 & 1/\sqrt{2} & -1/\sqrt{2} \end{bmatrix} \begin{bmatrix} s_1(t) \\ s_2(t) \\ s_3(t) \end{bmatrix}. \tag{5.6}$$

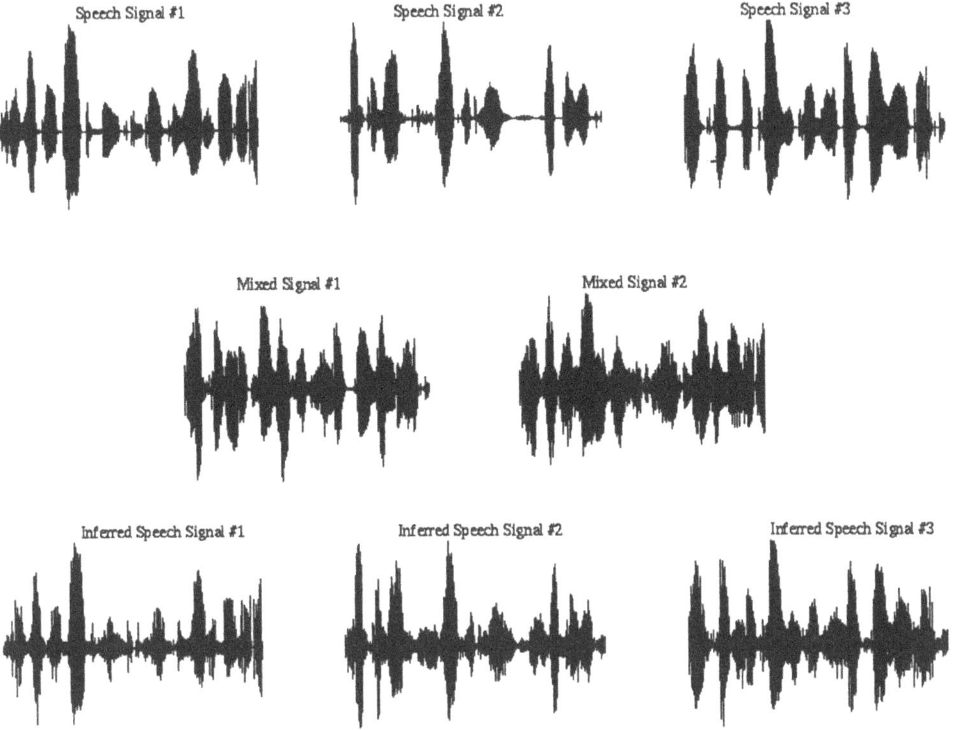

Figure 5.2. Demonstration of the separation of three speech signals from two mixtures. (Top row) The time course of 3 speech signals. (Middle row) Two observations of three mixed speech signals. (Bottom row) The inferred speech signals.

Figure 5.2 (middle) shows the time course of the two mixed speech signals. The 2-dimensional scatter plot (x_1 against x_2) in figure 5.3 (left) shows the three directions of the data. The three basis vectors of **A** were initially chosen randomly and were learned using eq.5.5. The learning process converged after 50 iterations. When more than 3 basis vectors were chosen, the amplitude of the redundant basis vectors converged to zero. The noise level l was set to 3 bits out of 8, i.e. the maximum amplitude of the noise signal was $2^3/2^8 \approx 3\%$ of the data range. Figure 5.3(right) shows the learned basis vectors. The sources were inferred using eq.5.3 and were recovered up to permutation and sign. Figure 5.2 (bottom) shows the three inferred speech signals after re-ordering and sign correction. The signal to noise ratio (SNR) for the separation was 20 dB, 17 dB and 21 dB respectively. Experiments with different speech signals and different mixing matrices yielded similar results. Although the temporal structure of the speech signal was not taken into consideration in the model, the separation quality was good [2].

The assumed noise-level, l, determines whether a data point should be considered as noise or as signal. A high noise-level ignores a wide range of data points around

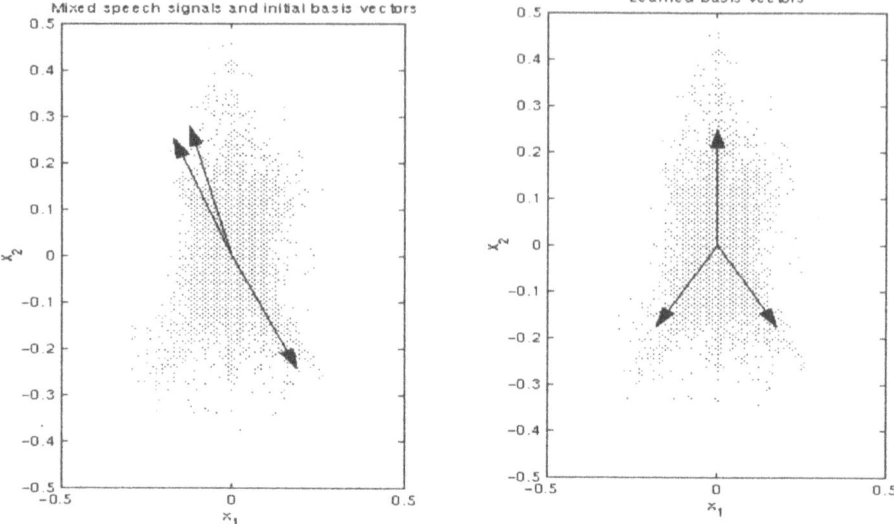

Figure 5.3. **Left**: Two-dimensional scatter plot of the two mixed signals. The three basis vectors were randomly initialized. **Right**: After convergence the learned basis functions are shown as arrows along the three speech signals. The learned basis vectors may be permuted and have a different sign.

zero and puts more weight on outliers when finding the basis vectors and when inferring the sources. This is significant in case of additive noise, where the appropriate noise-level may be adjusted to infer the sources. Figure 5.4 shows the SNR as a function of the noise-level l. Reasonably good SNR results were obtained for noise levels up to 6 bits (a maximum of the noise amplitude of 25%) and the performance degraded rapidly for a noise level of 7 bits or 8 bits.

The method of Lin et al. (1997) was also applied (see discussion) to this dataset. They inferred the sources by assuming that there was only one non-zero source at a given time sample . Using this method the SNR decreased by 4 dB, 2 dB and 7dB respectively .

5.2.2 Blind Separation of Speech and Music Signals

Experiments on music mixed with speech signals were performed as well. The distributions of music signals are in general less sparse than for speech signals and therefore the Laplacian density model may be less accurate. Three sources (the first two speech signals in the above example and one music signal) were mixed into two mixtures. Figure 5.5 shows the original signals, the mixtures and the inferred signals.

The SNR of the inferred signals using a noise-level of 3-bits were 17.3 dB and 16.8 dB for the speech signals and 14.2 dB for the music signal. The results were

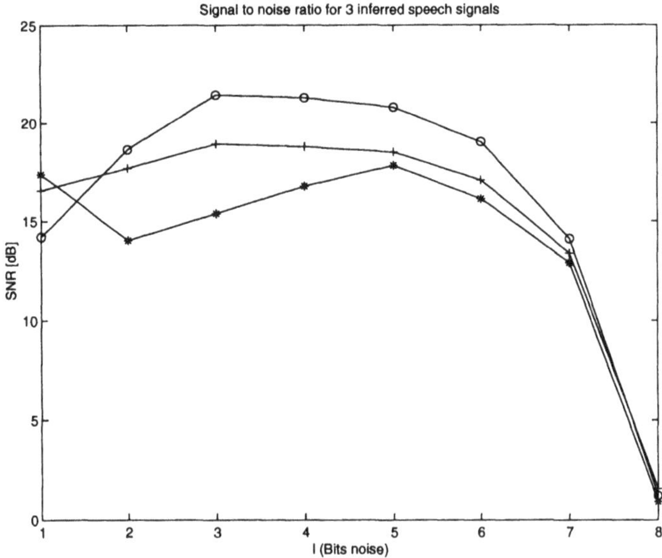

Figure 5.4. Signal to noise ratio (SNR) as a function of noise-level l. * = speech signal 1; + = speech signal 2 and o = speech signal 3.

comparable to those with the three speech signals despite the fact that the sparseness assumption on the sources was violated by the music source.

5.2.3 *Preliminary results with other mixtures*

Several speech mixing experiments were performed with varying number of sources and sensors. With two mixtures the proposed method was able to extract up to 4 mixed speech signals but the algorithm failed to find correct basis vectors when more than 4 sources were mixed into 2 observations. However, 5 speech signals were extracted from observations of 3 mixtures.

The formulation used here may also be used to unmix signals that were mixed with additive noise as assumed for the model in eq.5.1. Preliminary results indicate that overcomplete ICA can recover highly noisy mixtures and obtain a reasonable SNR. For noise-levels of 4 to 7 bits, the ICA algorithm used here recovered two mixed speech signals with additive Gaussian noise with a 5 dB to 10 dB improvement in SNR compared to the standard ICA (Bell and Sejnowski, 1995).

5.3 DISCUSSION

5.3.1 *Comparison to other methods*

The problem of separating more sources than observations has been treated several other methods. Pajunen (1997) and Hermann and Yang (1996) proposed methods for the special case of binary sources. Pajunen (1997) used a maximum likelihood

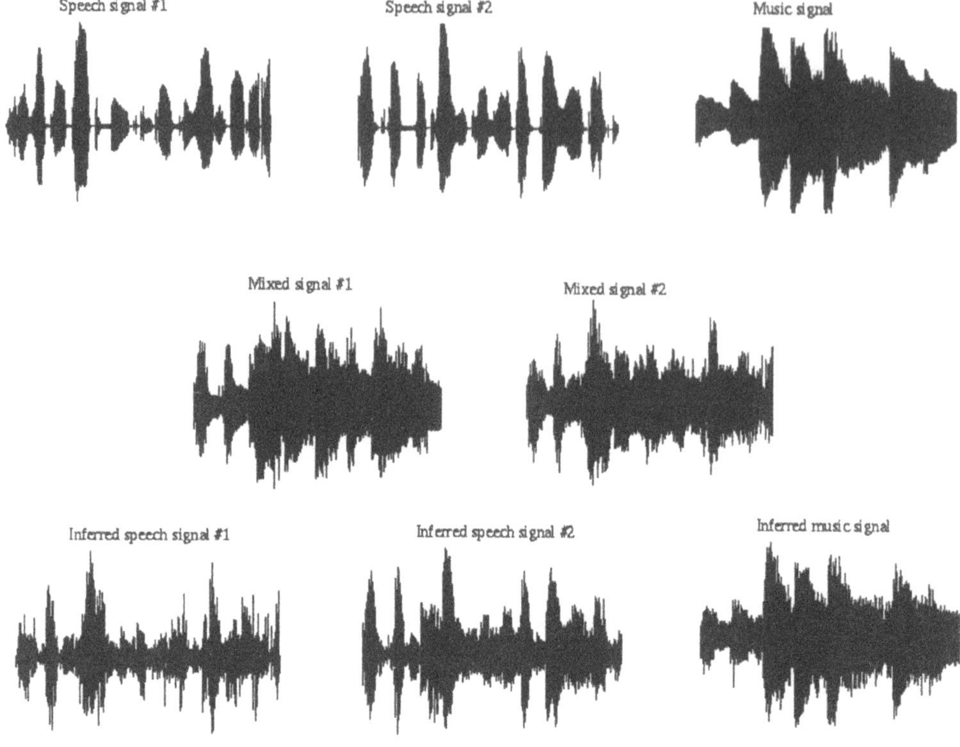

Figure 5.5. Blind separation of speech and music. (Top row) The time course of 2 speech signals and a music signal. (Middle row) These were mixed into two observations. (Bottom row) The inferred speech and music signals.

approach to reduce the problem to finding M clusters in the mixture space. Hermann and Yang (1996) applied self-organizing maps to find the clusters for binary sources. Lin et al. (1997) proposed a method for continuous signals by employing image analysis tools to detect geometric structure of the 2-dimensional mixture data locating the extremal density directions and thus finding the basis vectors. The sources were inferred by assuming that there was only one non-zero output at a given time, i.e. each data point was assigned to one source with the closest basis vector and all other sources were set to zero. In the presented experiments, this inference method gave inferior SNR for the speech separation example. The overcomplete ICA approach can be applied to continuous signals and is not restricted to binary sources. Furthermore, the probabilistic framework allows more flexible models which might lead to more accurate inferences.

Overcomplete representations can be learned in high-dimensional data space. For example, Lewicki and Olshausen (1998) used two times overcomplete basis to find 2×144 basis vectors for 12 by 12 patches of natural images.

5.3.2 Conclusions

The results presented here demonstrate that overcomplete representations can be used for blind source separation when there are more sources than mixtures. Reasonably good separations were obtained for two mixtures of 3 speech signals and for two mixtures of 2 speech signals and 1 music signal. Overcomplete representations reduce to ICA when the number of mixtures is equal to or greater than the number of sources. Currently investigations include the use of overcomplete representations of EEG data for artifact removal and for EEG signal detection with small numbers of sensors.

Notes

1. In most ICA formulations, the matrix $\mathbf{A}$ is restricted $M \geq N$, which is not imposed here.
2. The original, mixed and inferred signals are available in `http://www.cnl.salk.edu/~tewon/`.

6 FIRST STEPS TOWARDS NONLINEAR ICA

... and nothing shall be impossible unto you.
Matthew (17:20)

6.1 OVERVIEW

In many real world situations the linear assumption is an approximation of nonlinear phenomena. For several situations the linear assumption may lead to incorrect solutions. Therefore the goal in this chapter is to formulate an ICA framework that is able to separate nonlinear mixing models. Researchers have very recently started addressing the ICA formulation to nonlinear mixing models (Burel, 1992; Hermann and Yang, 1996; Lee et al., 1997c; Lin and Cowan, 1997; Pajunen, 1996; Taleb and Jutten, 1997; Yang et al., 1997) The proposed nonlinear ICA methods can be roughly divided into two classes of approaches. The first class of methods is an obvious extension to the linear ICA model where nonlinear mixing models are added to the linear model and the task is to find the inverse of the linear model as well as the inverse of the nonlinear model (Burel, 1992; Lee et al., 1997c; Taleb and Jutten, 1997; Yang et al., 1997). The nonlinearities are often parameterized allowing limited flexibility. More recently, Hochreiter and Schmidhuber (1998) have proposed low complexity coding and decoding approaches for nonlinear ICA. The second class of methods uses self-organizing-maps (SOM) to extract nonlinear features in the data (Hermann and Yang, 1996; Lin and Cowan, 1997; Pajunen, 1996) Their approach is more flexible and to some extent *parameter-free* which allows greater freedom of nonlinear representation.

In this chapter, the focus is on the first class of methods. A set of algorithms are proposed for the nonlinear mixing problem using parametric nonlinear functions. First, a model is presented where the mixing process is performed in two stages: a linear mixing followed by a nonlinear transfer function. A parametric sigmoidal nonlinearity and nonlinearities approximated by higher-order polynomials are suggested to solve the post-nonlinear problem. A similar approach was independently proposed by Taleb and Jutten (1997). They approximated the inverse transfer function by multilayer perceptrons (MLP) that were trained in an unsupervised manner. Those models may be justified for several biomedical signal analysis problems such as functional magnetic resonance imaging (fMRI) and electroencephalpgraphic (EEG) data analysis. It may also be used to account for intrinsic nonlinearities in a microphone that has been used in speech recording experiments. For these problems this model may be an appropriate representation of the actual physical phenomenon. The main drawback of this simple model is that the problem becomes inherently difficult and intractable when nonlinear mixing occurs between the cross-channels. A general framework is impossible and therefore a second simple nonlinear mixing model is proposed that takes the mixing into account up to second-order statistics. It can be shown that for certain cases, the mixed signals can be nonlinearly expanded the up to second-order. By applying a linear transformation onto the expanded mixtures the independent sources can be recovered under certain circumstances. Due to ambiguity introduced by the nonlinear mixing process a set of independent realizations are derived, in which one realization recovers the sources.

This chapter is organized as follows: In section 6.2, the first nonlinear model is presented and a set of learning rules is derived in section 6.3 based on the information maximization criterion. The learning rules are verified via simulation in section 6.4. In section 6.5 a second mixing model is proposed that is called a linearization method to nonlinear ICA. Other methods and future research are discussed in section 6.6.

6.2 A SIMPLE NONLINEAR MIXING MODEL

Figure 6.2 shows the mixing system which is divided into a linear mixing part and a nonlinear transfer part. Each channel i consists of an invertible nonlinear transfer function $f_i(t_i)$. The unmixing system is the inverse sequence of the mixing system. Figure 6.2 shows that first the nonlinear transfer function is inverted in each channel i with $h_i(x_i)$ and second the sources are unmixed by applying $\mathbf{W}$ to $\mathbf{z}$. The sources $\mathbf{s}$ are recovered if $h_i(x_i)$ and $\mathbf{W}$ are the inverse functions for $f_i(t_i)$ and $\mathbf{A}$ respectively.

This model uses the following signals: $\mathbf{s} = [s_1, s_2 \ldots, s_N]^T$, $\mathbf{t} = [t_1, t_2, \ldots, t_N]^T$, $\mathbf{x} = [x_1, x_2, \ldots, x_N]^T$, $\mathbf{z} = [z_1, z_2, \ldots, z_N]^T$, $\mathbf{u} = [u_1, u_2, \ldots, u_N]^T$, $\mathbf{f} = [f_1(t_1), f_2(t_2), \ldots, f_N(t_N)]^T$, $\mathbf{h} = [h_1(x_1), h_2(x_2), \ldots, h_N(x_N)]^T$. Furthermore, the signals are related by the fol-

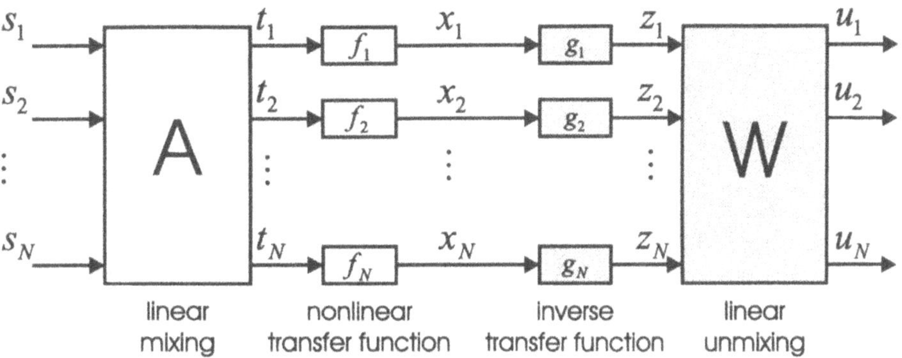

Figure 6.1. Mixing and unmixing model for nonlinear ICA: The mixing stage consists of a linear mixing matrix **A** and a nonlinear transfer function $\mathbf{f}(\mathbf{t})$. The unmixing stage consists of the inverse operation - the equalization of the nonlinear transfer function $\mathbf{g}(\mathbf{x})$ and the unmixing matrix **W**

lowing equations

$$\mathbf{t} = \mathbf{A} \cdot \mathbf{s} \tag{6.1}$$

$$\mathbf{x} = \mathbf{f}(\mathbf{t}) \tag{6.2}$$

$$\mathbf{z} = \mathbf{h}(\mathbf{x}) \tag{6.3}$$

$$\mathbf{u} = \mathbf{W} \cdot \mathbf{z} = \mathbf{W} \cdot \mathbf{h}[\mathbf{f}(\mathbf{A} \cdot \mathbf{s})]. \tag{6.4}$$

6.3 LEARNING ALGORITHM

Bell and Sejnowski (1995) have proposed an information-theoretic approach where they maximize the mutual information that an output $\mathbf{y}$ of a neural processor contains about its input $\mathbf{x}$. They have shown that for monotone and bounded mappings $g(\mathbf{u})$ and $\mathbf{u} = \mathbf{W}\mathbf{x}$, the mutual information between inputs and outputs can be maximized by maximizing the entropy of the outputs alone where the output pdf satisfies (see chapter 2)

$$p(\mathbf{y}) = \frac{p(\mathbf{x})}{|\mathbf{J}|}, \tag{6.5}$$

with $\mathbf{J}$ being the determinant of the Jacobian of the neural transfer function $g(\mathbf{u})$. This principle is adopted for the nonlinear mixing model in figure 6.2 and derive learning rules for the estimation of $\mathbf{W}$ and the parameters in $\mathbf{h}(.)$.

The joint entropy of the signals $\mathbf{y}$ is given by

$$H(\mathbf{y}) = -E\{\log p(\mathbf{y})\} = E\{\log |\mathbf{J}|\} - E\{\log p(\mathbf{x})\}. \tag{6.6}$$

Information maximization is performed by maximizing the first term with respect to the parameters of the unmixing functions. The goal is to learn the elements of the linear unmixing matrix $\mathbf{W}$ and the set of parameters for the nonlinearities $h_i(x_i)$. Using a gradient ascent algorithm the derivative of the entropy function with respect to w_{ij} and the parameters of the nonlinearity is

$$\Delta\mathbf{W} \propto \frac{\partial H(\mathbf{y})}{\partial\mathbf{W}} = \frac{\partial}{\partial\mathbf{W}} \log|\mathbf{J}|. \tag{6.7}$$

The second term in eq.6.6 is independent of all model parameters. Hence, the gradient of equation (6.7) is as follows

$$\frac{\partial}{\partial\mathbf{W}} \log|\mathbf{J}| = \frac{\partial}{\partial\mathbf{W}} \log|\det(\mathbf{W})| + \frac{\partial}{\partial\mathbf{W}} \log\left(\prod_{i=1}^{N} \frac{\partial y_i}{\partial u_i}\right) + \frac{\partial}{\partial\mathbf{W}} \log\left(\prod_{i=1}^{N} \frac{\partial h_i}{\partial x_i}\right). \tag{6.8}$$

Considering the set of parameters $\mathbf{W}$, a better way to maximize entropy in the feedforward and feedback system is not to follow the Euclidean gradient but the 'natural' gradient (Amari, 1997a)

$$\Delta\mathbf{W} \propto \frac{\partial H(\mathbf{y})}{\partial\mathbf{W}} \mathbf{W}^T\mathbf{W}. \tag{6.9}$$

This is an optimal rescaling of the entropy gradient. It simplifies the learning rule and speeds convergence considerably (see section 2.8.1).

6.3.1 Learning Rules for Sigmoidal Nonlinear Mixing

The infomax criterion holds for the model in figure 6.2 since independent variables cannot become dependent by passing them through an invertible nonlinearity. Hence, the mutual information before and after the nonlinear stage is not affected.

For the derivation of the learning rule for the w_{ij} the last term of eq.6.8 is not considered. Therefore, the learning rule for $\mathbf{W}$ is

$$\Delta\mathbf{W} \propto (\mathbf{W}^T)^{-1} + (1-2\mathbf{y})\mathbf{h}^T(\mathbf{x}). \tag{6.10}$$

Considering the natural gradient from equation (6.9) it follows

$$\Delta\mathbf{W} \propto \mathbf{W} + (1-2\mathbf{y})\mathbf{u}^T\mathbf{W}. \tag{6.11}$$

Although this learning rule is derived for super-Gaussian sources the rule may be extended to the separation of sub-Gaussian sources as presented in chapter 2 (see eq.2.56). For the reason of simplicity, the learning rules for a logistic activation function is derived which is well suited to separate super-Gaussian sources. Consider the parametric form of the nonlinear transfer function f_i

$$f_i(t_i) = \delta_i \left(1 - \frac{2}{1+\exp(-\sigma_i t_i)}\right), \tag{6.12}$$

where δ_i denotes the scaling and σ_i the slope of the transfer function. For this case, $h_i(x_i)$ provides the inverse function by

$$h_i(x_i) = -2r_i \operatorname{arctanh}(d_i x_i), \tag{6.13}$$

whereas the equalities $r_i = 1/\sigma_i$ and $d = 1/\delta_i$ hold in the ideal case. Gradient ascent on the entropy function to learn the parameters d and r gives

$$\Delta\mathbf{r} \propto \frac{\partial H(\mathbf{y})}{\partial \mathbf{r}} \quad \text{and} \quad \Delta\mathbf{d} \propto \frac{\partial H(\mathbf{y})}{\partial \mathbf{d}} \tag{6.14}$$

$$\frac{\partial H(\mathbf{y})}{\partial \mathbf{r}} = \frac{\partial}{\partial \mathbf{r}} \log|\det(\mathbf{W})| + \frac{\partial}{\partial \mathbf{r}} \log\left(\prod_{i=1}^{N} \frac{\partial y_i}{\partial u_i}\right) + \frac{\partial}{\partial \mathbf{r}} \log\left(\prod_{i=1}^{N} \frac{\partial h_i}{\partial x_i}\right) \tag{6.15}$$

and

$$\frac{\partial H(\mathbf{y})}{\partial \mathbf{d}} = \frac{\partial}{\partial \mathbf{d}} \log|\det(\mathbf{W})| + \frac{\partial}{\partial \mathbf{d}} \log\left(\prod_{i=1}^{N} \frac{\partial y_i}{\partial u_i}\right) + \frac{\partial}{\partial \mathbf{d}} \log\left(\prod_{i=1}^{N} \frac{\partial h_i}{\partial x_i}\right). \tag{6.16}$$

The term $(\det \mathbf{W})$ in the equations (6.15) and (6.16) is independent of $\mathbf{r}$ and $\mathbf{d}$. Hence,

$$\begin{aligned} \frac{\partial}{\partial r_j} \log\left(\prod_{i=1}^{N} \frac{\partial y_i}{\partial u_i}\right) &= \sum_{i=1}^{N} (1 - 2y_i) w_{ij} \frac{\partial}{\partial r_j} h_j(x_j) \\ &= -2\text{arctanh}(d_j x_j) \sum_{i=1}^{N} (1 - 2y_i) w_{ij} \end{aligned} \tag{6.17}$$

and

$$\begin{aligned} \frac{\partial}{\partial d_j} \log\left(\prod_{i=1}^{N} \frac{\partial y_i}{\partial u_i}\right) &= \sum_{i=1}^{N} (1 - 2y_i) w_{ij} \frac{\partial}{\partial d_j} h_j(x_j) \\ &= -2r_j \frac{x_j}{1 - d_j^2 x_j^2} \sum_{i=1}^{N} (1 - 2y_i) w_{ij}. \end{aligned} \tag{6.18}$$

The third term is

$$\frac{\partial}{\partial r_j} \log\left(\prod_{i=1}^{N} \frac{\partial h_i}{\partial x_i}\right) = \frac{1}{r_j} \tag{6.19}$$

and

$$\frac{\partial}{\partial d_j} \log\left(\prod_{i=1}^{N} \frac{\partial h_i}{\partial x_i}\right) = 1 + 2d_j^2 x_j^2 (1 - d_j^2 x_j^2)^{-1}. \tag{6.20}$$

6.3.2 Learning Rules for Flexible Nonlinearities

A weakness of the sigmoidal nonlinearity is that the learning rules can be successfully applied to only those problems which fit to the parametric structure of a sigmoid.

The microphone sensor nonlinearities may be approximated by a sigmoid function. However, in certain situations where the a priori knowledge about the mixing model is not given a more flexible nonlinear transfer function is required. Assume that a nonlinearity may be approximated by polynomials of Qth-order. This nonlinear stage may be described as

$$f_j(t_j) = \sum_{k=1}^{Q} f_{jk} \cdot t_j^{k-1}. \tag{6.21}$$

The inverse $h_j(x_j)$ of the function in eq.6.21 results in an expression which is in general complicated. Therefore it is assumed that the inverse may be approximated by P^{th}-order polynomials. Then, the inverse is

$$h_j(x_j) = \sum_{k=1}^{P} h_{jk} \cdot x_j^{k-1}. \tag{6.22}$$

In the same manner, a gradient ascent on the entropy function is performed to learn the parameters h_{jk}

$$\Delta h_{jk} \propto \frac{\partial H(\mathbf{y})}{\partial h_{jk}}. \tag{6.23}$$

Performing this operation on eq.1.4, the learning rule for finding h_{jk} is the sum of the following two terms

$$\begin{aligned} \frac{\partial}{\partial h_{jk}} \log\left(\prod_{i=1}^{N} \frac{\partial y_i}{\partial u_i}\right) &= \sum_{i=1}^{N}(1 - 2y_i)w_{ij}\frac{\partial}{\partial h_{jk}}h_j(x_j) \\ &= x_j^{k-1}\sum_{i=1}^{N}(1 - 2y_i)w_{ij} \end{aligned}$$

and

$$\frac{\partial}{\partial h_{jk}} \log\left(\prod_{i=1}^{N} \frac{\partial h_i}{\partial x_i}\right) = \frac{x_j^{k-2}(k-1)}{\sum_{m=1}^{P}(m-1) \cdot h_{j,m}x_j^{m-2}}. \tag{6.24}$$

6.4 SIMULATION RESULTS

Sigmoidal Nonlinearities

To verify the validity of the model and the convergence of the learning rules, several experiments were performed with the architecture shown in figure 6.2. Figure 6.4 shows the result of the mixing and unmixing system. Two independent white noise sources with super-Gaussian distributions were generated artificially and are shown

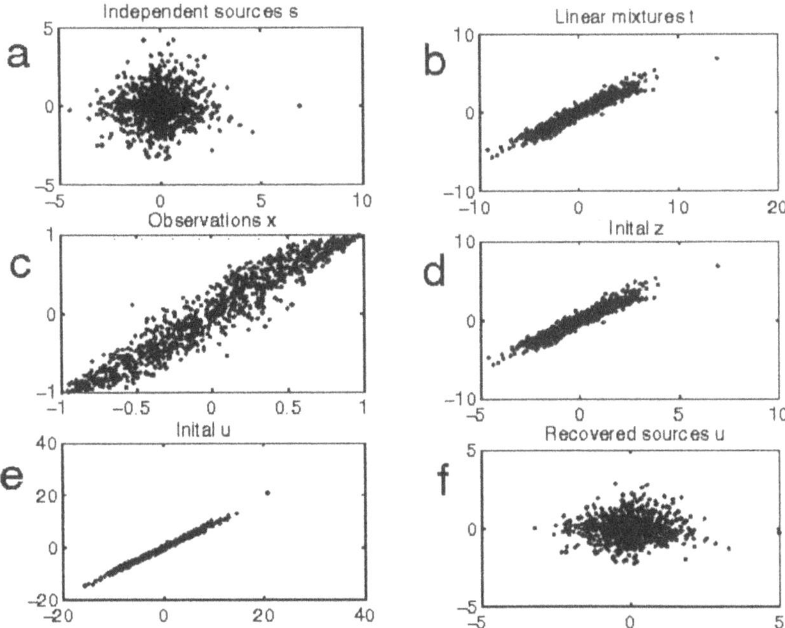

Figure 6.2. Mixing and unmixing simulation for super-Gaussian sources, scatter-plot: (a) independent sources (b) linearly mixed sources (c) nonlinear mixing (d) initially unmixed nonlinearity (e) initially separated signals $\mathbf{u}$ (f) finally separated signals $\mathbf{u}$.

in a scatter plot in figure 6.4 (a). The sources were first mixed linearly (b) and then transferred nonlinearly (c) with a logistic function f_i.

$$\begin{bmatrix} t_1 \\ t_2 \end{bmatrix} = \begin{bmatrix} 1 & 2 \\ 1 & 1 \end{bmatrix} \begin{bmatrix} s_1 \\ s_2 \end{bmatrix}, \tag{6.25}$$

$$\begin{aligned} x_1 &= f_1(0.5t_1) \\ x_2 &= f_2(t_2). \end{aligned} \tag{6.26}$$

The unmixing results in figure 6.4 (d) and (e) when the nonlinearities are initialized identically and the unmixing matrix **W** is chosen randomly. The algorithm converged after presenting 500 samples and the unmixed signals are shown in figure 6.4(f). The Signal to Noise Ratio (SNR) for the observed mixed signals x_1, x_2 are -6.3 dB and -6.1 dB respectively. For the unmixed signals u_1, u_2 the SNR is increased to 8.9 dB and 8.0 dB respectively.

Flexible Nonlinearities

As for the logistic nonlinearities, several experiments were performed with the architecture shown in figure 6.2 to verify the learning rules for flexible nonlinearities. For the linear and nonlinear mixing stage the same mixing matrix as in eq.6.26 was used. The independent sources were white noise signals with sub-Gaussian distributions. A scatter plot of the sources is depicted in figure 6.2 (a). The unmixing **W** and the coefficients h_{jk} of the nonlinearity forming polynomials are chosen randomly with $Q = P$. Figure 6.2 (f) shows the results after presenting 1000 samples. Better separation results were obtained when the order of the inverse nonlinearity $h(.)$ was higher than the order of the nonlinearity $f(\mathbf{t})$. The stability of the polynomial nonlinearity is highly dependent on the initial value of the coefficients. Its form was imposed to approximate an invertible nonlinearity. When applied to the sigmoidal nonlinearity in the previous section the polynomial approximation gave similar results.

In figure 6.3 (a) and (b) the time course of a sinusoid signal and a white noise signal with super-Gaussian distribution are shown. The signals were mixed linearly (c,d) and transformed by a nonlinear transfer function $f(t)$ where $f(t)$ was an invertible 5th-order polynomial function. The inverse nonlinearity $h(x)$ was approximated by a 8th-order polynomial function. The time course of the recovered signals are shown in figure 6.3 (e) and (f).

6.5 A LINEARIZATION APPROACH TO NONLINEAR ICA

This section presents a more general approach to nonlinear mixing. Instead of considering a linear mixing and a nonlinear transfer function model the nonlinear mixing is formulated as follows

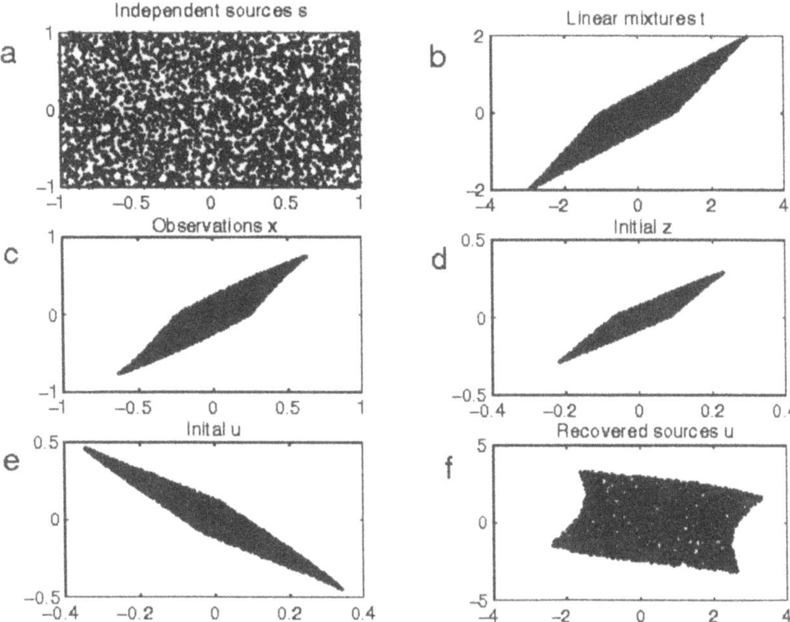

Figure 6.3. Simulation results using flexible nonlinearities: (a) independent sources (b) linearly mixed sources (c) nonlinear transferring (d) initially unmixed nonlinearity (e) initially separated signals u (f) finally separated signals u_1 and u_2.

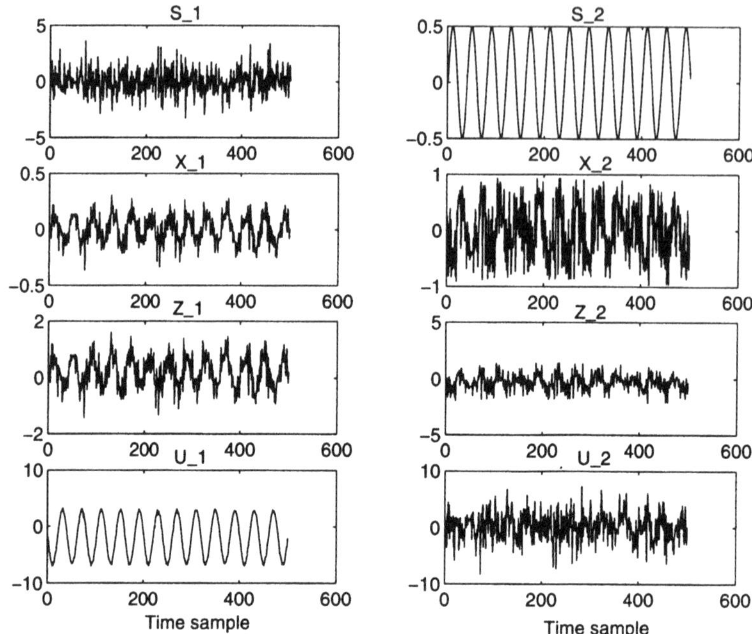

Figure 6.4. Mixing and unmixing simulation using flexible nonlinearities. (Top) Time course of the original signals, (2nd row) the linear mixtures, (3rd row) the mixtures after the nonlinear transfer function and (*bottom*) the recovered sources.

$$\begin{aligned} x_1 &= f_1(s_1, \cdots, s_N) \\ x_2 &= f_2(s_1, \cdots, s_N) \\ x_N &= f_N(s_1, \cdots, s_N). \end{aligned} \tag{6.27}$$

The observations $x_1, x_2, \cdots, x_N$ are nonlinear functions $f_1, f_2, \cdots, f_N$ of the independent sources $s_1, s_2, \cdots, s_N$.

The nonlinearity f_i can be written in a Taylor series expansion as

$$f_i(s_1, \cdots, s_N) = \sum_{k=0}^{n} \frac{f^k(\mathbf{s}_0)}{k!}(\mathbf{s} - \mathbf{s}_0)^k + R_n(\mathbf{s}), \tag{6.28}$$

where the Taylor expansion is about $\mathbf{s}_0$ and $R_n(\mathbf{s})$ is called the Lagrange's form of the remainder in the Taylor series expansion of $f_i(.)$. The Taylor series expansion can be approximated up to second-order around zero-mean sources giving

$$f_i(s_1, \cdots, s_N) = \sum_{k=0}^{2} \mathbf{a}_k \mathbf{s}^k + R_2(\mathbf{s}) \tag{6.29}$$

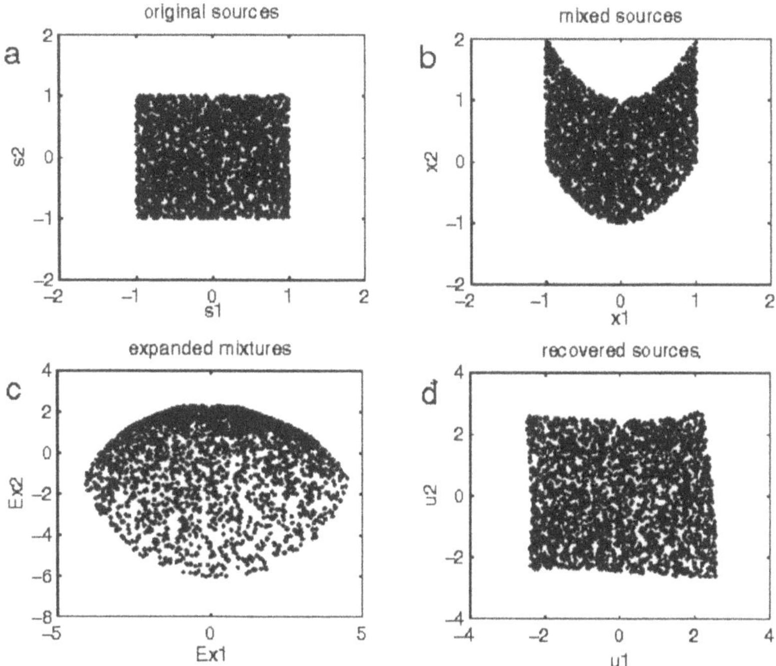

Figure 6.5. Example for the linearization approach to nonlinear ICA. (*top-left*) original sources s_1 and s_2, (top-right) nonlinearly mixed sources x_1 and x_2, (*bottom-left*) expanded observation x_{e1} and x_{e2} and (*bottom-right*) recovered sources u_1 and u_2.

where $\mathbf{a}_k$ are the characteristic coefficients of $f(.)$ up to k^{th}-order.

For simplicity, consider two sources and two sensors. Furthermore, assume that the sources can become mixed up to second-order only. Then, x_1 and x_2 reduces to the following form

$$\begin{aligned} x_1 &= a_{11}s_1^{(0,1)}s_1^{(0,1)} + a_{12}s_2^{(0,1)}s_2^{(0,1)} + a_{13}s_1^{(0,1)}s_2^{(0,1)} \\ x_2 &= a_{21}s_1^{(0,1)}s_1^{(0,1)} + a_{22}s_2^{(0,1)}s_2^{(0,1)} + a_{23}s_1^{(0,1)}s_2^{(0,1)}, \end{aligned} \tag{6.30}$$

where $s_i^{(0,1)}$ denotes that the signal can be taken either to the power of zero (which leads to one) or one (which gives the signal). Eq.6.30 approximates the nonlinear observation of two independent sources up to second-order mixing.

The two signals x_1 and x_2 can be nonlinearly expanded using a polynomial series truncated at second order. This gives the following 5 signals

$$\begin{aligned} x_{e1} &= x_1 \\ x_{e2} &= x_2 \\ x_{e3} &= x_1^2 \\ x_{e4} &= x_2^2 \\ x_{e5} &= x_1x_2. \end{aligned} \tag{6.31}$$

Given the expanded observation, the idea is now to apply a linear transformation so that the expanded observation becomes linearly independent (Wiskott, personal communication). In fact, this principle is related to some ideas in support vector machines used for classification (Vapnik et al., 1997).

Assume a simple example

$$\begin{aligned} x_1 &= a_{11}s_1 \\ x_2 &= a_{22}s_2 + a_{12}s_1^2. \end{aligned} \tag{6.32}$$

The second-order expanded terms are

$$\begin{aligned} x_{e1} &= x_1 = a_{11}s_1 \\ x_{e2} &= x_2 = a_{22}s_2 + a_{12}s_1^2 \\ x_{e3} &= x_1^2 = a_{11}^2s_1^2 \\ x_{e4} &= x_2^2 = a_{22}^2s_2^2 + 2a_{22}a_{11}^2s_2s_1^2 + a_{11}^4s_1^4 \\ x_{e5} &= x_1x_2 = a_{11}s_1a_{22}s_2 + a_{11}^3s_1^3. \end{aligned} \tag{6.33}$$

A linear transformation $\mathbf{W}$ is sought that makes the expanded terms independent ($\mathbf{u} = \mathbf{W}\mathbf{x}_e$). One possible solution for this example is when the weights to u_1 are all zero except for the direct weight x_{e1}. This gives $u_1 = x_{e1} = a_{11}s_1$. The other source can be recovered when $u_2 = x_{e2} - x_{e3} = a_{22}s_2$. However, this is not the only solution. The following equations give a valid solution as well

$$\begin{aligned} u_1 &= x_{e3} = a_{11}^2s_1^2 \\ u_2 &= x_{e2} - x_{e3} = a_{22}s_2. \end{aligned} \tag{6.34}$$

Clearly, u_1 and u_2 are independent but the original sources are not recovered. Hence, an independence criteria is not sufficient to recover the original source. The ICA solution is satisfied when the outputs u_i are independent which is the case for eq.6.33 and eq.6.34.

Several simulations were performed using this example and changing the coefficients a_{ij} in eq.6.32. The initial signals u_1 and u_2 are shown in figure 6.32 (*bottom-left*). After five passes through the 3000 points data the sources are recovered. In figure 6.5 the scatter-plot is shown for all possible combinations between the outputs u_i, u_j. Figure 6.5 (c) shows the other valid solution in eq.6.34. Except of (a) and (c) all other scatter-plots show dependencies between the sources. Since the recovered sources can be permuted, the valid solutions have to be verified by another measure of independence between the outputs u_i that decides which combination of two outputs is most independent.

6.6 DISCUSSIONS

6.6.1 *Other approaches to nonlinear ICA*

Other approaches to the nonlinear ICA problem using self-organizing maps (SOM) were proposed by Hermann and Yang (1996); Lin and Cowan (1997); Pajunen (1996). SOM constitute a class of vector quantizers that impose a prescribed topological order on the reference (input) vectors (Kohonen, 1989). When the network structure is equivalent to the topology of the sources, i.e. if the common Cartesian product of one dimensional spaces is used, then for certain conditions the SOM represent the inverse of the mixing transformation. This is achieved by mapping the mixed observation signals onto a regular output grid where each coordinate of the SOM output represents one source. An advantage of SOMs is that they can provide a non-parametric approach to the nonlinear ICA framework. In fact, SOMs can be used in certain cases for under-determined problems (Lin et al., 1997) and noisy input signals. However, the nature of SOM-based approaches allows to recover sources only in a discretized manner or via interpolation. The introduced quantization error can be reduced by increasing the size of the network. The number of sources and the size of the network have a great impact on the computational cost (exponentially with the number of sources). Therefore, SOMs are more suited for low dimensional problems with low precision.

Similar approaches to our presented methods using the linear mixing and nonlinear transfer function was developed by Taleb and Jutten (1997) Other methods considering cross channel nonlinearities were proposed by Burel (1992) and Yang et al. (1997). In their approaches, the nonlinear observation $\mathbf{f}(\mathbf{x})$ was linearly mixed by a second mixing matrix $\mathbf{W_2}$. The learning rules can be derived to find $\mathbf{W_1}$, $\mathbf{W_2}$ and one nonlinearity $\mathbf{h}(\mathbf{x})$. In contrast to their approach, subject of our future interest is to find nonlinear cross-channels which can be parameterized independently from the channel transfer functions.

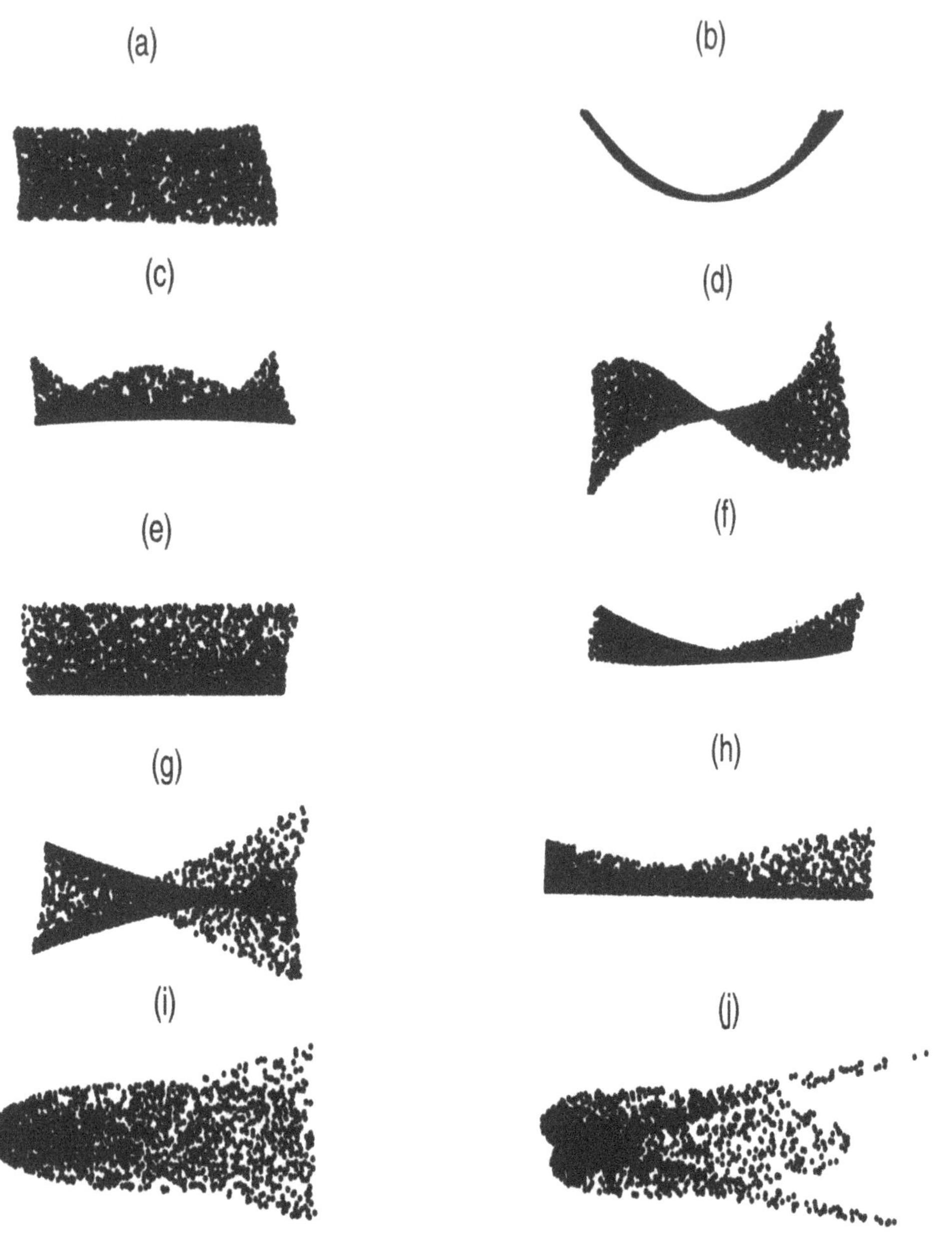

Figure 6.6. All possible combinations between the five outputs u_i. Only (a) and (c) represent valid solutions.

6.6.2 Conclusions and future research

Two nonlinear ICA approach were presented. In the first model, a set of learning rules was derived for the nonlinear blind source separation problem based on the information maximization criterion where the mixing model was divided into a linear mixing part and a nonlinear transfer channel. The proposed algorithms were focused on a parametric sigmoidal nonlinearity and higher order polynomials. Simulation results were performed to verify the learning rules. In the second model, an approach that can incorporate a more general mixing model was proposed by using a polynomial expansion and linearization method.

II Independent Component Analysis: Applications

7 BIOMEDICAL APPLICATIONS OF ICA

I am not built for academic writings.
Action is my domain.
Gandhi

7.1 OVERVIEW

This chapter deals with the application of ICA to biomedical data. In biomedical recordings multiple sensors are used to record some physiological phenomena. Often these sensors are located close to each other, so that they simultaneously record signals that are highly correlated with each other. For example, in electroencephalographic (EEG) recordings the sensors are placed at the scalp within a few centimeters of each other. Therefore, the sensors not only record brain activity transmitted by volume conduction from a few dynamic neocortical processes but also artifactual signals, such as noise independent of brain processes, that overlap with neural brain activity that and may be present in all sensors. In this case, a useful tool for EEG researchers would be a method for segregating neural brain activity from artifactual noise signals, or even more interesting, a method that can segregate overlapping neural activities into independent components. Given some major assumptions about EEG signals: (1) that they sum approximately linearly, and (2) are temporally independent, ICA may be an appropriate tool to blindly separate overlapping EEG signals and artifacts into independent components.

Makeig et al. (1996) have first applied spatial ICA (using instantaneous mixing only) to the analysis of EEG data and event-related potential (ERP) data using the original infomax algorithm (Bell and Sejnowski, 1995). Their results indicate that EEG recordings can be decomposed into overlapping EEG phenomena, including alpha and theta bursts. Furthermore, they indicate that ICA is able to segregate obvious artifactual components. Independently, Vigario et al. (1996) and Karhunen et al. (1997b) report similar findings for EEG recordings using a fixed-point algorithm (Hyvaerinen and Oja, 1997a; Hyvaerinen, 1997) related to minimizing the 4th-order cumulants.

A general assumption made in EEG and ERP analysis is that the signals are generated by distinct neural sources, i.e. cortical patches, neural networks or all assemblies. Furthermore, the temporal distribution of EEG activations are sparse assuming each macroscopic source is active during only a small part of the experiment (Makeig et al., 1997). For this case, the original infomax algorithm is well suited to decompose overlapping EEG activities. However, some artifacts in the data, e.g. line noise, are nearly constantly active, exhibiting a distribution that is sub-Gaussian. For these cases, Jung et al. (1998a) showed that the extended infomax (Lee et al., 1998b) algorithm is able to linearly decompose EEG artifacts into independent components having both sub- and super-Gaussian distributions.

EEG recordings are one example of using ICA for biomedical signal processing. There may be many more applications in which ICA shows superior analysis qualities than traditional statistical methods such as PCA. Very recently, McKeown et al. (1998b) have used the extended ICA algorithm to investigate task-related human brain activity in functional Magnetic Resonance Imaging (fMRI) data. fMRI techniques have recently been shown to be a powerful method of studying neocortical dynamic functions. In fMRI data there are thousands of data channels which record brain activity simultaneously at different brain locations (voxels). Therefore, this technique can show for each location (layer or brain slice and coordinate) the time course of its activation. However, the slight spread of brain activity onto many adjacent voxels lead to a smeared version of the original signals. Here as well, artifacts occur during the recording. Under these circumstances ICA may again be an appropriate tool to discover independent components in fMRI data and to segregate artifacts from the data.

This chapter summarizes the main results obtained using the extended infomax algorithm on EEG and fMRI data. First, it is demonstrated how the algorithm can isolate artifacts from overlapping EEG signals such as alpha and theta bursts. The algorithm can effectively separate 60 Hz line noise from all channels into one channel. Second, on another set of EEG data, the extended infomax algorithm can effectively isolate eye blinks, eye movements, line noise, cardiac contamination and muscle noise from EEG recordings. Those artifacts can be removed from the ICA representation which, when projected back into the data space, is then free of the artifacts. Other independent components which can be separately projected back onto the scalp show patterns of activity related to the neurophysiology of the subject. Third, the effects of using the extended infomax algorithm to analyze fMRI data is summarized. By imposing independent spatial fMRI maps, sparse and localized maps are found showing voxels that have either consistently or transiently task-related brain activations. These activations are difficult to detect in raw fMRI data. Compared to other methods, such as PCA, ICA detects more regions of task-related brain activations, corroborating findings of Positron Emission Tomography (PET) studies.

The following literature is suggested for details about the physiological interpretation and comparison to other methods of ERP, EEG and fMRI analysis: A general book on EEG analysis from an engineering perspective, Nunez (1981); a good review of ERP studies given by Hillyard and Picton (1980); a book on MRI and fMRI Toga and Mazziotta (1996). Regarding the application of ICA to EEG, ERP and fMRI data, Makeig et al. (1997), Jung et al. (1998d) and McKeown et al. (1998b) explain the methods and results in detail.

This chapter is organized as follows: In section 7.2 the use of extended infomax for EEG analysis is demonstrated. Section 7.3 describes the methods and results of isolating artifacts in EEG. Section 7.4 describes how extended infomax can be used to find consistently and transiently task-related brain activations in fMRI data. Finally, in section 7.5 the limitations and future research in ICA for biomedical data analysis is discussed.

7.2 ICA OF ELECTROENCEPHALOGRAPHIC DATA

Electroencephalographic (EEG) recordings of brain electrical activity measure changes in potential difference between pairs of points on the human scalp. Scalp recordings also include artifacts such as line noise, eye movements, blinks and cardiac signals (ECG) which can present serious problems for analyzing and interpreting EEG recordings (Berg and Scherg, 1991).

The ICA algorithm appears to be very effective for performing source separation in domains where, (1) the mixing medium is linear and propagation delays are negligible, (2) the time courses of the sources are independent, and (3) the number of sources is greater or equal the number of sensors, i.e. if N sensors are used the ICA algorithm can separate a maximum of N sources. In the case of EEG signals, volume conduction is thought to be linear and instantaneous, hence assumption (1) is satisfied. Assumption (2) is also reasonable because the sources of eye and muscle activity, line noise, and cardiac signals are not generally time locked to the sources of EEG activity which is thought to reflect activity of cortical neurons. Assumption (3) is questionable since the effective number of statistically-independent signals contributing to the scalp EEG is unknown. However, numerical simulations have confirmed that the ICA algorithm can accurately identify the time courses of activation and the scalp topographies of relatively large and temporally-independent sources from simulated scalp recordings, even in the presence of a large number of low-level and temporally-independent source activities (Ghahremani et al., 1996; Makeig et al., 1997).

For EEG analysis, the rows of the input matrix $\mathbf{x}$ are the EEG signals recorded at different electrodes, the rows of the output data matrix $\mathbf{u} = \mathbf{W}\mathbf{x}$ are time courses of activation of the ICA components, and the columns of the inverse matrix $\mathbf{W}^{-1}$ give

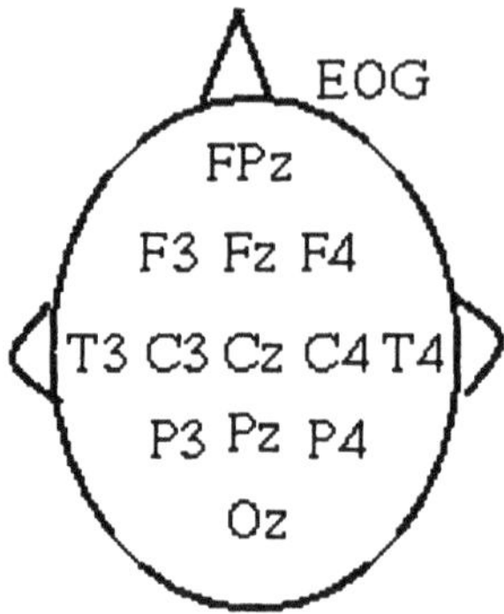

Figure 7.1. The location of the electrodes for 14 channels according to the International 10-20 System.

the projection strengths of the respective components onto the scalp sensors. The scalp topographies of the components allows to examine their biological plausibility (e.g., eye activity should project mainly to frontal sites). For artifactual removal, the unwanted components can be eliminated by setting the corresponding source components to zero in $\hat{\mathbf{u}}$ (estimated independent components) and projecting back to a new dataset $\hat{\mathbf{x}}$ that is free of the artifacts. The inverse projection is performed in two steps because the basis functions of $\mathbf{W}$ are not orthogonal. The solution $\mathbf{W}$ can be separated into an orthogonal matrix $\mathbf{W}_{\text{orth}}$ and a whitening matrix $\mathbf{V}$ as follows

$$\mathbf{W} = \mathbf{V}\mathbf{W}_{\text{orth}}. \tag{7.1}$$

To project back into $\hat{\mathbf{x}}$, first project $\hat{\mathbf{u}}$ onto $\mathbf{W}_{\text{orth}}^{-1}$ ($\hat{\mathbf{u}}_{\text{tmp}} = \mathbf{W}_{\text{orth}}^{-1}\hat{\mathbf{u}}$) giving $\hat{\mathbf{u}}_{\text{tmp}}$ and second project as $\hat{\mathbf{x}} = \mathbf{V}^{-1}\hat{\mathbf{u}}_{\text{tmp}}$.

7.2.1 Simple examples of applying ICA to EEG data

EEG data sets were analyzed that were collected to develop a method of objectively monitoring the alertness of operators listening for auditory signals (Makeig and Inlow, 1993; Jung et al., 1996). During half-hour session, the subject was asked to push one button whenever they detected an auditory target stimulus. EEG was collected from 14 electrodes located at sites of the International 10-20 System at a sampling rate of 312.5 Hz. Figure 7.1 shows the ordering of the electrodes. The extended infomax algorithm was applied to the 14 channels of 10 seconds of data with the following parameters: learning rate fixed at 0.0005, 100 passes with block size of 100 (3125 weight updates). The power spectrum was computed for each channel and the power in a band around 60 Hz was used to compute relative power for each channel and each separated component.

Figure 7.2 shows the time course of 14 channels EEG and figure 7.3 shows the independent components found by the extended infomax algorithm. Several observations on the ICA components in figure 7.3 and its power spectrum are of interest

- Alpha bursts (about 11 Hz) are detected in components 1 and 5. Alpha band activity (8-12 Hz) occurs most often when eyes are closed and the subject is relaxed. Most subjects have more than one alpha rhythm, with somewhat different frequencies and scalp patterns.

- Theta bursts (about 7 Hz) are detected in components 4, 6 and 9. Theta-band rhythms (4-8 Hz) also have different patterns and may occur under distinctly different conditions. For example, during drowsiness increases in theta activity accompany transient losses of awareness or microsleeps (Makeig and Inlow, 1993), while frontal theta bursts occur during intense concentration.

- An eye blink is detected in component 2, at 8 sec.

- 60 Hz line noise in component 3 (see figure 7.4,bottom). The extended infomax algorithm effectively concentrates the line noise present in nearly all the channels into one ICA component.

In addition, figure 7.4 (top) shows the ratio of power near 60 Hz distributed over the channels. In the EEG data, the 60 Hz is dominant in channels 4 and 14. However, all channels exhibit 60 Hz line noise. Figure 7.4 (middle) shows that the original infomax algorithm cannot concentrate the line noise into one component. In contrast, extended infomax (figure 7.4, bottom) picks up the 60 Hz line noise component and concentrates it mainly in one sub-Gaussian component, channel 3.

Figure 7.5 shows another EEG data set with 23 channels including 2 EOG (electrooculogram) channels. The eye blinks near 5 sec and 7 sec contaminated all of the channels. Figure 7.6 shows the ICA components without normalizing the components with respect to their contribution to the raw data. ICA component 1 in figure 7.6 contained the pure eye blink signal. Small periodic muscle spiking at the temporal sites (T3 and T4) was extracted into ICA component 14.

Experiments with several different EEG data sets confirmed that the separation of artifactual signals was highly reliable. In particular, severe line noise signals could always be decomposed into one or two components with sub-Gaussian distributions. Jung et al. (1998a) show further that eye movement also can be extracted. In these cases, the originally proposed algorithm by Bell and Sejnowski (1995) could not clearly isolate the artifacts.

7.3 EEG ARTIFACT REMOVAL USING EXTENDED INFOMAX

Severe contamination of EEG activity by eye movements, blinks, and muscle, heart and line noise presents a serious problem for EEG interpretation and analysis. Eye movements, muscle noise, heart signals, and line noise often produce large and distracting artifacts in EEG recordings. Rejecting EEG segments with artifacts larger than an arbitrarily preset value is the most commonly used method for dealing with artifacts in research settings. However, when limited data are available, or blinks and muscle movements occur too frequently, as in some patient groups, the amount of data lost to artifact rejection may be intolerable. Here, the results in Jung et al.

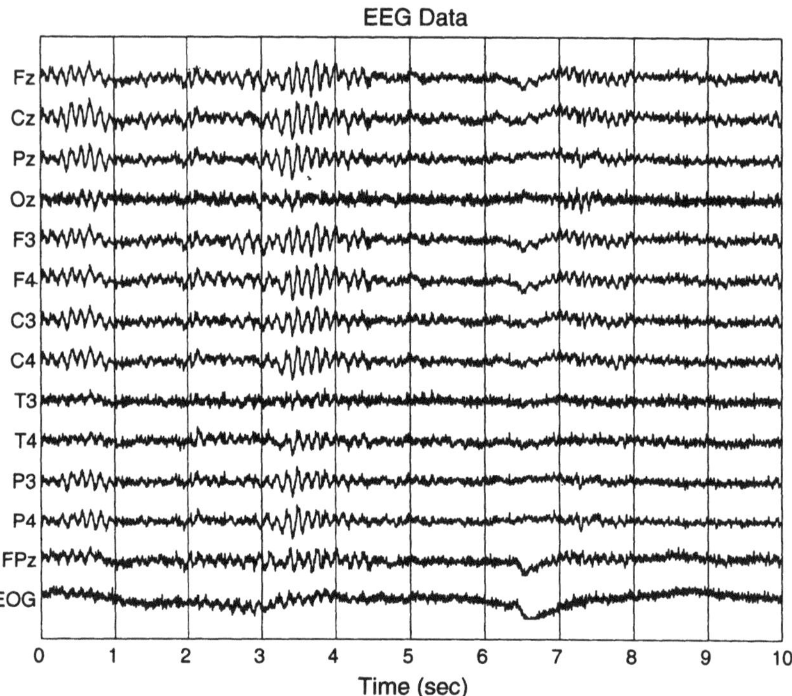

Figure 7.2. A 10-sec portion of the EEG time series with prominent alpha rhythms (8-21 Hz). The location of the recording electrode from the scalp is indicated on the left of each trace. The electrooculogram (EOG) recording is taken from the temples.

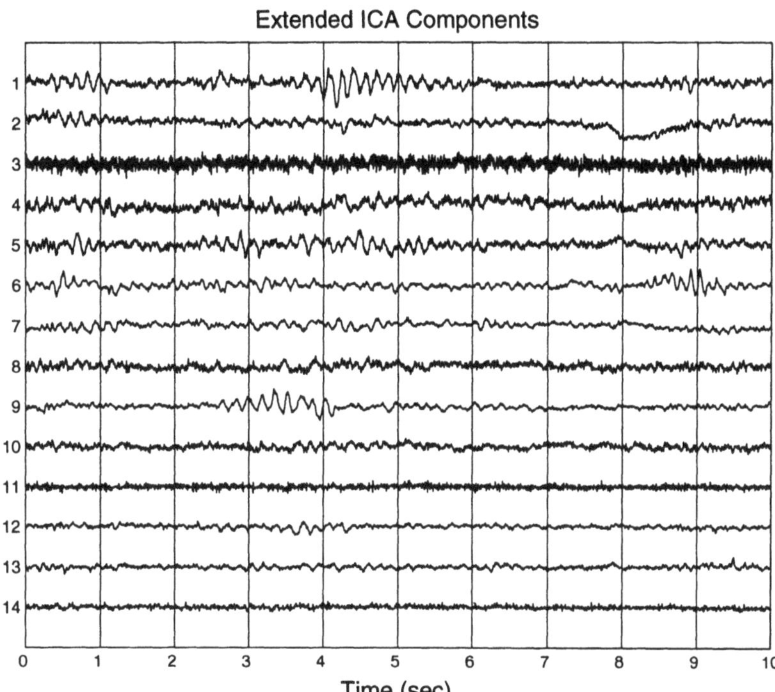

Figure 7.3. The 14 ICA components extracted from the EEG data in figure 7.2. Components 3, 4, 7, 8 and 10 have sub-Gaussian distributions in the others have super-Gaussian distributions. There is an eye movement artifact at 8 seconds. Line noise is concentrated in component 3. The prominent rhythms in components 1,4,5,6 and 9 have different time courses and scalp distributions.

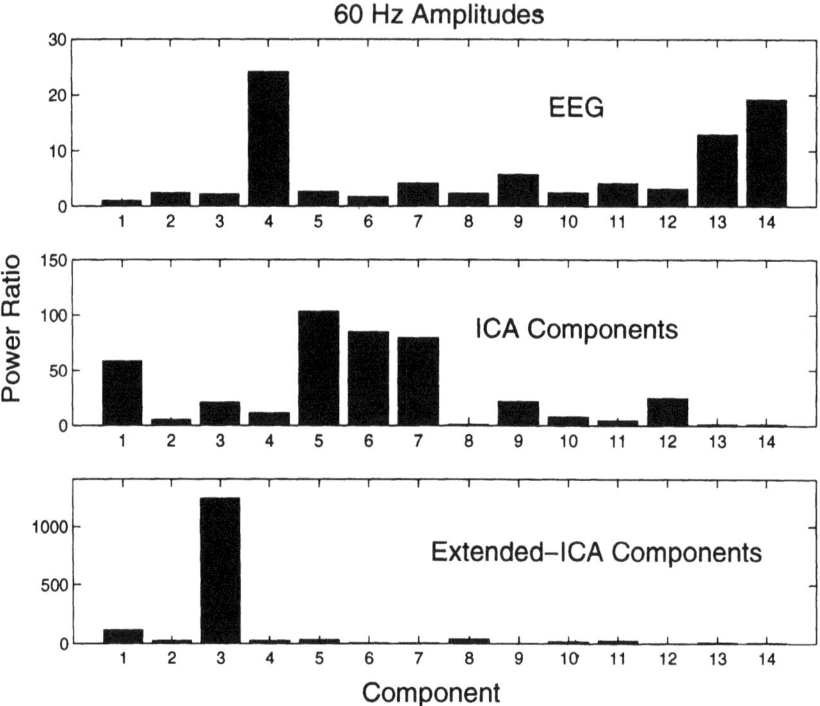

Figure 7.4. Top: Ratio of power near 60 Hz over 14 channels for EEG data in figure 7.2. Middle: Ratio of power near 60 Hz for the 14 infomax ICA components. Bottom: Ratio of power near 60 Hz for the 14 extended infomax ICA components in figure 7.3. Note the difference in scale by a factor of 10 between the original infomax and the extended infomax.

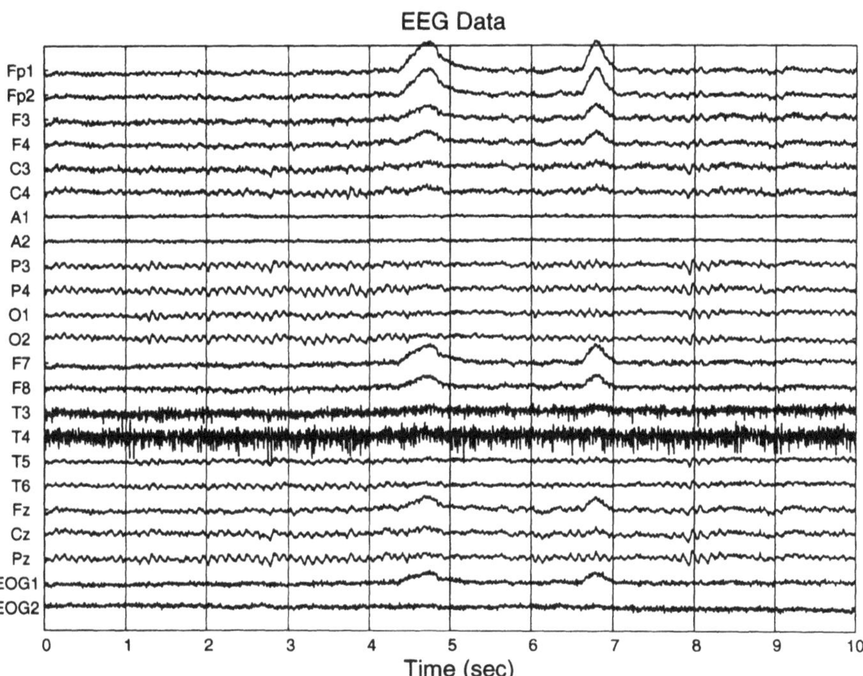

Figure 7.5. EEG data set with 23 channels including 2 EOG channels. At around 4-5 sec and 6-7 sec artifacts from severe eye blinks contaminate the data set.

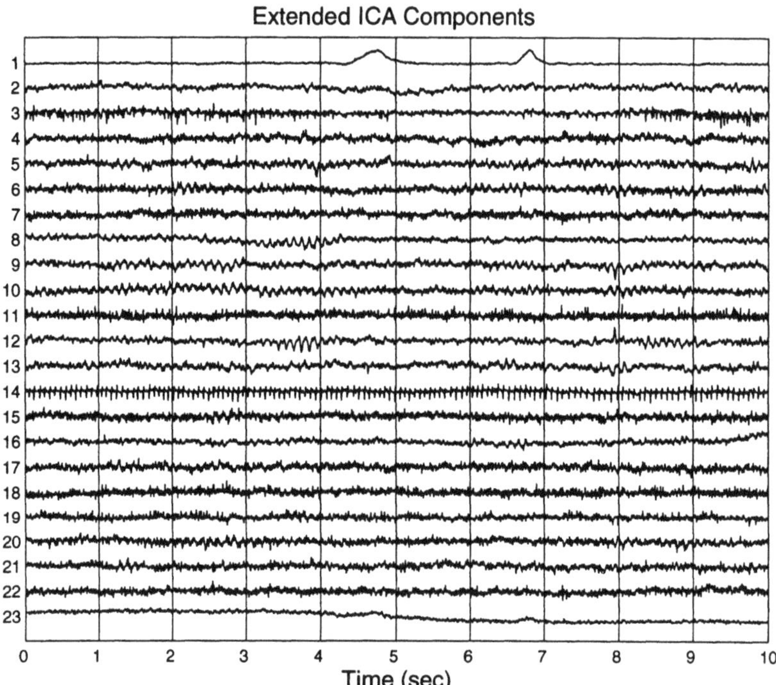

Figure 7.6. Extended infomax ICA components derived from the EEG recordings in figure 7.5. The eye blinks are clearly concentrated in component 1. Component 14 contains the steady state signal.

(1998a, 1998c) and Jung et al. (1998b) are summarized in which several other artifacts are extracted using the extended infomax algorithm.

7.3.1 Methods and Materials

An EEG data set used in the analysis was collected from 19 scalp electrodes placed according to the International 10-20 System and from 2 EOG placements, all referred to the left mastoid. The sampling rate was 256 Hz. ICA decomposition was performed on 5-sec EEG epochs from each data set using Matlab 4.2c on a DEC 2100A 5/300 processor. The learning batch size was 90, and initial learning rate was 0.001. Learning rate was gradually reduced to 5×10^{-6} during 80 training iterations requiring 6.6 min of computer time. Figure 7.7 (*left*) shows a 5-sec portion of the recorded EEG time series and its ICA component activations in figure 7.7 (*middle*). Figure 7.8 shows the time course of five artifacts and figure 7.9 shows the scalp map of four artifactual signals. These maps are generated as follows: 1) The column of the inverse weight matrix $\mathbf{W}$ accounts for the amount of intensity that a component contributes to each electrode. 2) The intensity of an electrode is mapped into a circle. 3) A colormap is assigned to display the magnitude of the component at each electrode. 4) Corresponding to the electrode's position, a two dimensional grid of data points can be generated by interpolating between the electrodes giving the scalp map. To obtain 'corrected' EEG signals shown in figure 7.7 (*right*) the back projection method in eq.7.1 is used to remove the artifactual signals.

The artifactual signals are discussed below in more detail:

- **Eye movement artifacts**

  Eye movement artifacts at 0.5, 2.0 and 4.7 sec (*left*) are detected and isolated to ICA component 2 (*middle left*), even though the training data contains no EOG reference channel. The scalp map of the component captures the spread of EOG activity to frontal sites. Component 5 represents horizontal eye movements, After eliminating these two components and projecting the remaining components onto the scalp channels, the 'corrected' EEG data (*right*) are free of these artifacts.

- **Muscle artifacts**

  Left temporal muscle activity in the data is concentrated in ICA component 3 (Fig. 7.7, *top middle*). The ICA component 3 reveals the presence of small periodic muscle spiking in left frontal channels (e.g., F4) which is highly obscured in the original data.

- **Cardiac contamination and line noise**

  Line noise has a sub-Gaussian distribution and so could not be clearly isolated by the original infomax algorithm. By contrast, the extended infomax algorithm effectively concentrates the line noise present in nearly all the channels into ICA component 4. The widespread cardiac contamination in the EEG data (*left*) is concentrated in ICA component 11.

After eliminating these five artifactual components, the 'corrected' EEG data (*right*) are free of these artifacts.

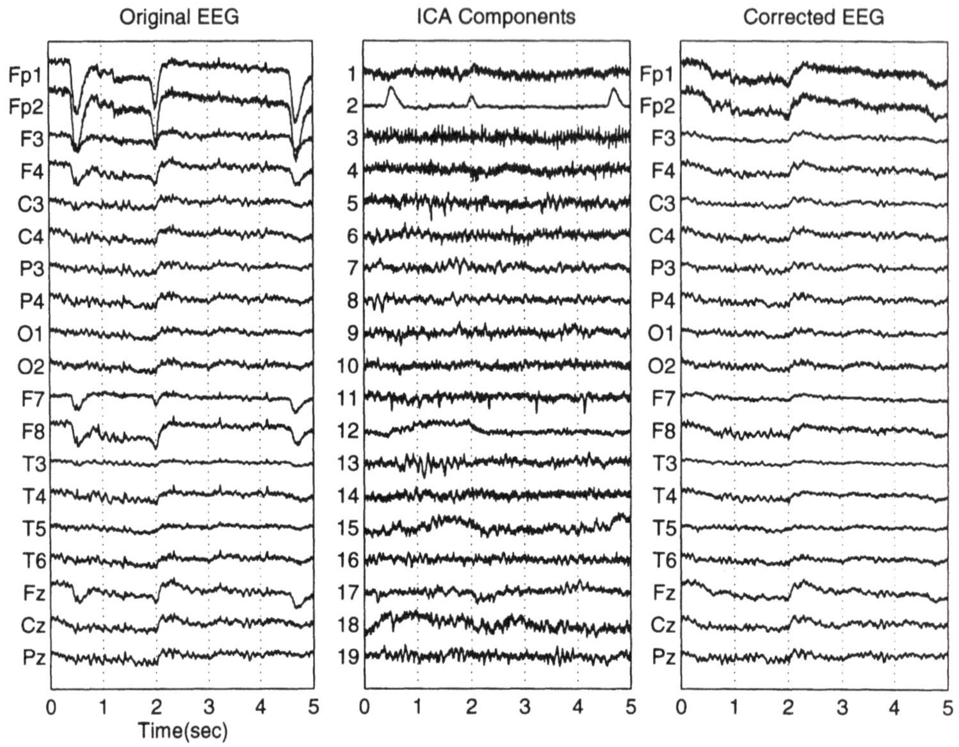

Figure 7.7. Artifact removal using the extended infomax algorithm. A 5-sec portion of the EEG time series (*left*), ICA components accounting for eye movements, cardiac signals, and line noise sources (*center*), and the EEG signals 'corrected' for artifacts by removing the five components (*right*).

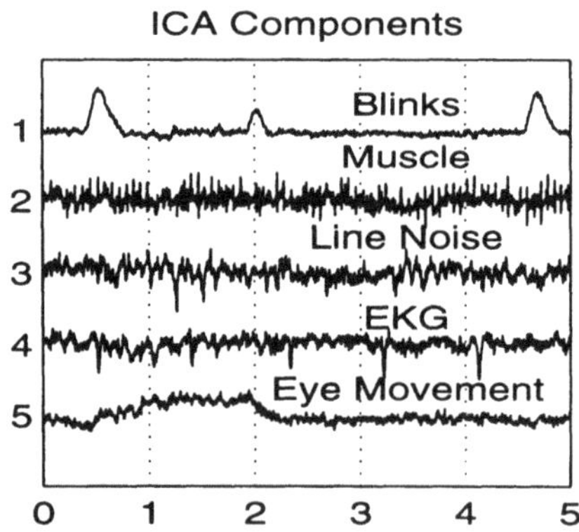

Figure 7.8. Time course of the five artifactual components only. ICA components accounting for eye blinks, muscle noise, cardiac noise signals, and slow eye movement.

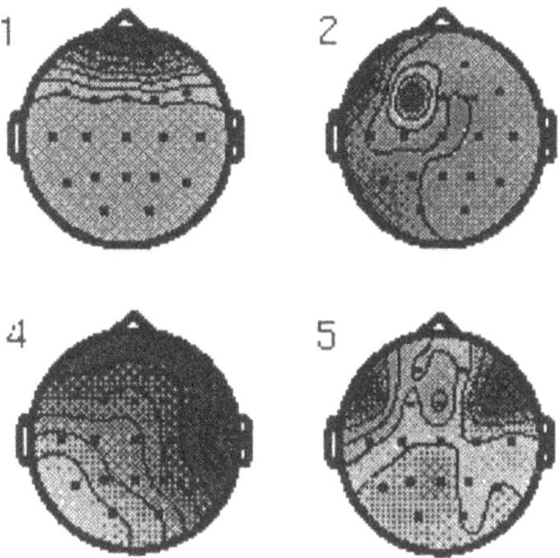

Figure 7.9. Scalp map of 4 artifactual components. Maps accounting for (1:top-left) eye blinks, (2:top-right) muscle noise, (3:bottom-left) cardiac noise signal and (4:bottom-right) show eye movement.

7.3.2 Discussion of EEG applications

ICA appears to be a generally applicable and effective method for removing a wide variety of artifacts from EEG records. There are several advantages of the method: (1) ICA is computationally efficient. (2) ICA is generally applicable to removal of a wide variety of EEG artifacts. It simultaneously separates both the EEG and its artifacts into independent components based on the statistics of the data, without relying on the availability of 'clean' reference channels. This avoids the problem of mutual contamination between regressing and regressed channels. (3) No arbitrary thresholds (variable across sessions) are needed to determine when regression should be performed. (4) Separate analyses are not required to remove different classes of artifacts. Once the training is complete, artifact-free EEG records in all channels can then be derived by simultaneously eliminating the contributions of various identified artifactual sources in the EEG record. However, because ICA is a statistical method, its results may not be meaningful when the amount of data or number of channels are insufficient. Future work includes determining the data length and number of input channels needed to remove artifacts of various types, and comparing the performance of the ICA method to that of other approaches.

7.4 FUNCTIONAL MAGNETIC RESONANCE IMAGING ANALYSIS

Functional Magnetic Resonance Imaging [1] (fMRI) [2] allows humans to be monitored during the performance of psychomotor tasks with moderately high temporal ($\sim$ 2 sec) and spatial ($\sim$ 5mm^2) resolution. The different paramagnetic susceptibilities of oxygenerated and deoxy-hemoglobin provide the basis for the noninvasive Blood Oxygenation Level Dependent (BOLD [3]) contrast measure most often used in current fMRI studies. Although blood volume and BOLD-signal responses to local neural activity changes are relatively slow, with rise times of about 5 to 8 sec, hemodynamic responses to stimulus and task onsets are generally reproducible between trials and sessions.

Current echo-planar technology allows to acquire MRI images up to 20 times a second. However due to the hemodynamic delay and due to changes arising from machine noise, subtle subject movements and heart and breathing rhythms it is very difficult to detect task-related activations. The non task-related signals may even account for most of the observed BOLD signal variance (Kwong et al., 1992).

Traditional statistical method to enhance fMRI data are based on subtraction and correlation averaged over a number of task-block cycles. Averaging over task-blocks however, reduces the sensibility of fMRI analysis to changes in brain activation occurring in one portion of a trial. Such transiently task-related (TTR) activations may arise from changes in subject performance, variations in subject arousal and attention or effort, or changes produced by learning. Therefore, it is desirable not to average over the fMRI data and to use a method that can find TTR activations.

McKeown et al. (1998b) have used ICA to find components that are transiently time-locked to the behavioral experiment as well as consistently task-related (CTR) activations. Their results suggest that ICA can be used to explore a wide range of psychological and neuro-cognitive processes occurring during relatively unstructured fMRI experiments.

7.4.1 fMRI Methods

A subject participated in two 6 min trials of a *Stroop color-naming task.* Each trial consists of five 40 sec control blocks alternating with four experimental task blocks. In the control blocks, the subject was simply required to covertly name the color of a displayed rectangle (red, blue or green) while in the *Stroop*-task blocks, the subject was required to name the discordant color of the script used to print a color name. For example, if the word 'green' was presented in blue script, the subject was to covertly 'say' the word 'blue' without speaking it. This experiment follows a pattern of alternating task blocks which represent our reference function. The goal is to find brain activations that are highly correlated with the reference function indicating that neural activity in the detected area may be required for this task.

A 1.5T GE Signa MRI system was used to monitor brain activity using the BOLD contrast. Ten 64 × 64 echo planar, gradient recalled (TR = 2500 ms, TE = 40 ms) axial images (5 mm think, 1 mm inter-slice gap) with a 24 cm field of view were collected at 2.5 sec sampling intervals corresponding to 146 images for each slice. Ten slices through the brain are recorded simultaneously from 10.000 voxels. So, at each time point 10.000 data points are acquired

The main assumption that is made to perform ICA on fMRI data is that the maps (slices of fMRI images) are independent of each other. A justification for this stems from observations made from different studies (fMRI, PET, EEG) suggesting that

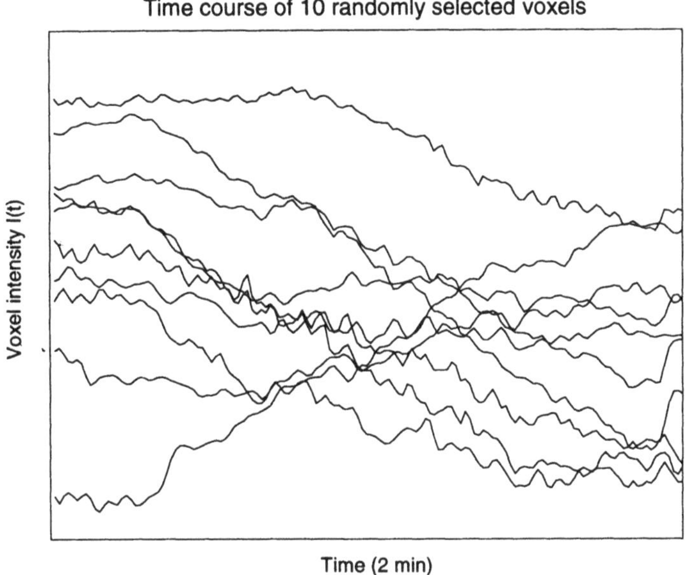

Figure 7.10. Time course of 10 randomly selected voxels recorded near the visual cortex.

brain activations are sparsely localized. The raw fMRI data however, shows smeared activations and inconsistency during different trials. ICA is applied to a data matrix $\mathbf{x}$ where the columns are the time-course of one voxel and the rows represent 10.000 voxels. For this data set a 146×146 weight matrix $\mathbf{W}$ is learned using the extended infomax learning algorithm in eq.2.56 with the following parameters: learning rate fixed at 0.0001, momentum term, block size of 100 and 300 passes through the data. Learning took about 4 hours with Matlab 4.2c on a DEC 2100A 5/300 processor.

7.4.2 fMRI Results

Figure 7.10 shows the raw fMRI data by selecting 10 voxels from the visual cortex. The signals correlated with the Stroop task reference function [4] are difficult to detect. In MRI, noise signals may significantly dominate so that the reference function is not detectable. Performing ICA transforms the data into independent maps. Voxels are found with time-courses that match exactly the time-courses of the reference function. Figure 7.11 shows the time-courses of the same voxels which are now highly correlated [5] with the reference function meaning that these brain areas are significantly task related. Figure 7.12 shows the map of one slice of the ICA-transformed fMRI data. Voxels that are correlated with the reference function are localized in the visual cortex (bottom of the slice image) and the left frontal cortex. These results suggest first that the neural activity in the visual cortex is active during the Stroop experiment. This is not surprising since the Stroop task is a visual task and the

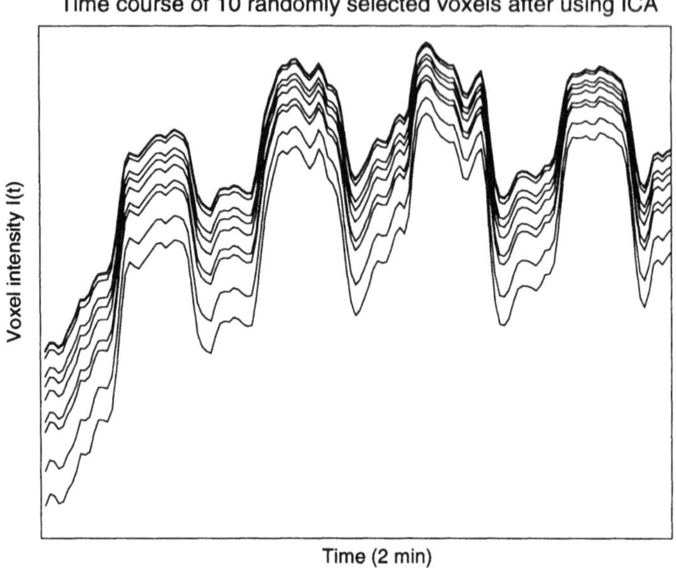

Figure 7.11. The same voxels show activations corresponding to the experiment after ICA transformation.

difference between the alternating experiments are well captured consistently active in the visual cortex. However, the results also suggest that there are frontal neural activities which most likely indicate neural computations performed by the brain during the Stroop task. These results confirm previous Positron Emission Tomography (PET) studies which reported occipital and medial frontal activations. Frontal activations have recently been linked to visual-spatial attention (Nobre et al., 1997), language processing (Binder, 1997) and working memory (Manoach et al., 1997) which may be involved in *Stroop*-task performance.

The same experiments were analyzed using other techniques. Figure 7.13 shows the results of PCA, 4th-order cumulants and extended infomax. Clearly, extended infomax is able to extract a time-course signal closely matching the Stroop task reference function.

7.4.3 Discussions and Conclusions on fMRI

The assumption about spatial independence of fMRI components used in the analysis may not be obvious although it matches neurophysiological observations. In fact, the idea was first introduced to make the algorithm computationally feasible (McKeown, personal communication). However, a temporal ICA formulation can be applied as follows: Given the time-course signals from the voxels, assume that each voxel records independent signals such as signals from neural activity and noise artifacts. Apply

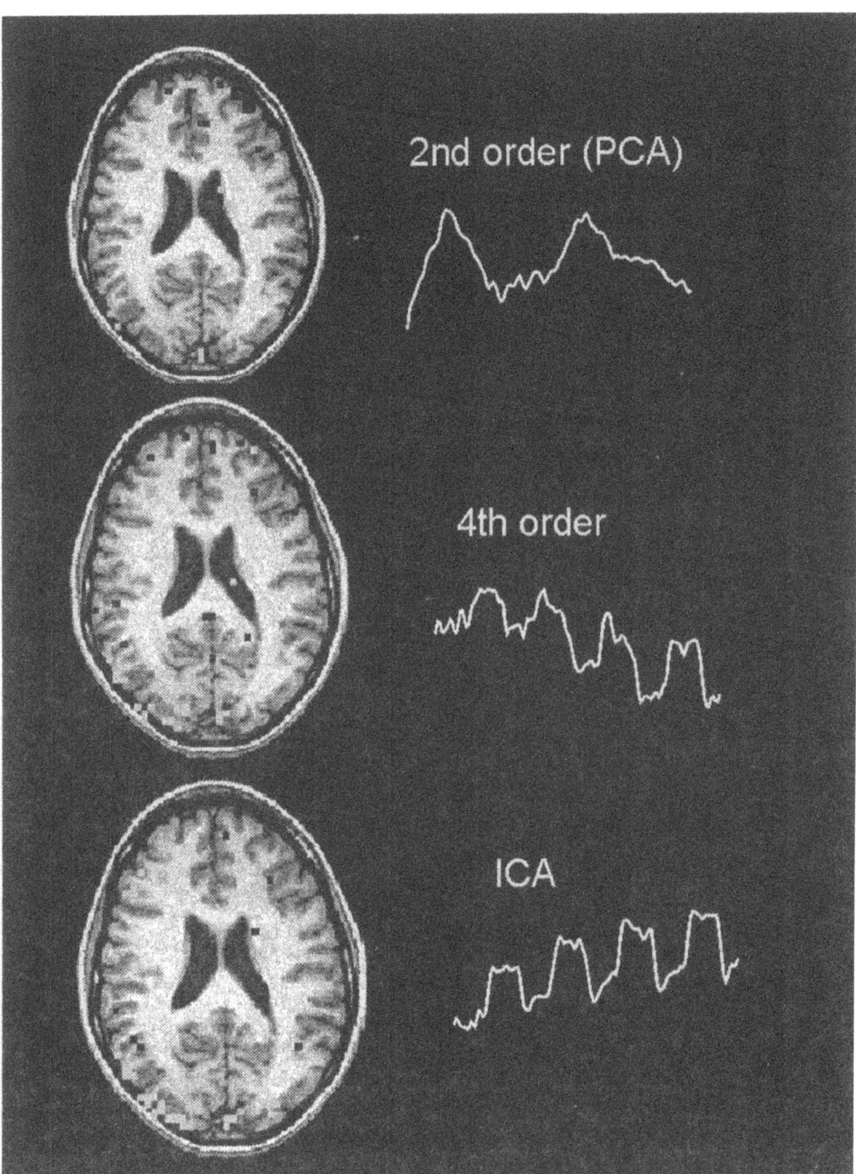

Figure 7.12. Brain maps of activations using different separation techniques. (top) Results of PCA analysis, (middle) results using 4th-order cumulants (Comon, 1994), and (bottom) results from the extended infomax algorithm.

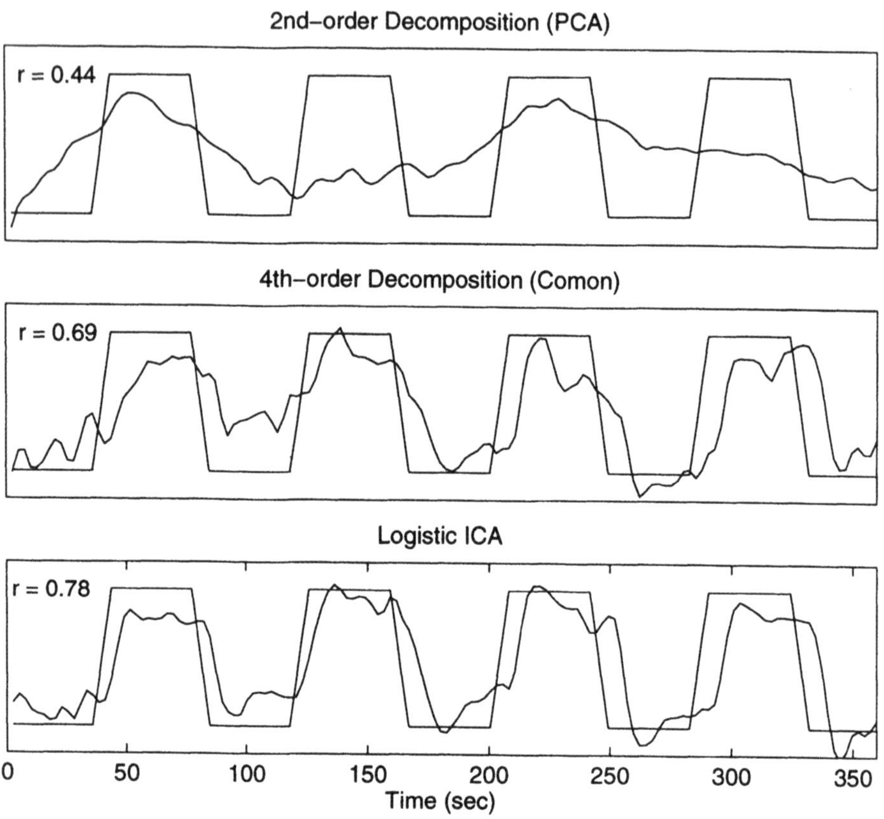

Figure 7.13. Comparison of different methods with respect to the Stroop task reference function. (top) Results from the PCA analysis; (middle) results from using 4th-order cumulants (Comon, 1994); (bottom) results using the extended infomax algorithm. (r denotes the correlation between the reference function and the detected component signal.)

the extended infomax algorithm to separate the mixed observations into independent components. In this data set, there are 10.000 sensors (voxels) and only 146 time points. Due to computational complexity (finding a 10.000×10.000 weight matrix) the following process is used: Define brain regions which are strongly active in accord with the task-related reference function as a region of interest (ROI) (McKeown et al., 1998a), i.e. a region consisting of 2000 voxels with high correlations between the voxel time-courses and the reference function. Use PCA to compress the data into a 10-by-146 point data set by projecting onto the 10 largest eigenvectors. Initial results using temporal ICA on the fMRI ROI data showed that this method can find activations in the frontal as well as in the visual area. Activations in the visual area are due to the visual nature of the task. However, frontal activations suggest which areas in the frontal cortex might account for brain cognitive or executive activity during the tasks. In addition to these activations, McKeown et al. (1998a) also find transiently activated time-courses correlated to the Stroop reference function during one or two of the alternating test blocks. This may shed light on learning mechanisms in which some areas may be activated only when given a new task.

Several questions about ICA decomposition still need to be addressed: ICA is a linear method and is still able to give good results even though most researchers assume that the hemodynamic response function is nonlinear and therefore the sensors should sum signals in a nonlinear manner. Furthermore, the response function also includes time-delays which are not considered in these experiments.

However, ICA seems to be sensitive to both transiently and consistently task-related brain activations. The method gives highly reproducible results and is consistent across different trials and different subjects. It may be also used to isolate artifactual components from fMRI data.

7.5 CONCLUSIONS AND FUTURE RESEARCH

The application of ICA to EEG and fMRI data analysis was presented. Given EEG data, ICA has been shown to separate overlapping EEG activities as well as interfering artifactual signals. Given fMRI data, ICA gives useful results for investigating task-related human brain activity. Spatial ICA on fMRI data gives sparsely distributed independent maps identifying local areas whose activations correspond to the time course of the reference function. In addition to obvious activity in the visual cortex during the Stroop task, the algorithm also detected consistently and transiently task-related brain activations in the frontal cortex, suggesting which parts of this brain area may contribute to the subject's computational or cognitive efforts during the task.

The presented methods for EEG and fMRI analysis are potential applications which may be considered as important software and hardware tools in the near future. An online-version of the algorithm mounted on a DSP chip might be useful for detecting sources during an experiment, allowing clinicians to make more reliable decisions based on the immediate ICA results.

There are several issues that researchers have just begun to address:

1. Nonlinear mixing

2. Nonstationary behavior

3. Blind deconvolution, time delays

4. Reliability given only a small data set.

Issue (1) is in particular of interest for fMRI where several models for the hemodynamic response function exist and this may be included as a priori knowledge for nonlinear mixed ICA. Issue (2) is now being addressed with different techniques and is being tested on EEG data (see chapter 9). Issue (3) may not apply for EEG but is of significance for fMRI data. Although there are methods for convolved and time-delayed sources as proposed in chapter 4, applications to multichannel biomedical data have still several problems. One problem is the number of filters that increases quadratically with the number of sensors. To give a reliable filter estimate, it is crucial to ensure a sufficient amount of training data which is in many cases limited. E.g., in fMRI 146 time-points are sampled, whereas in speech processing it is simple to record 100000 time samples. A statistical preprocessing technique for ICA to estimate the weight parameters given only a small data set (4) is presented by Koehler et al. (1997) combining a statistical method with infomax. This method gives a more reliable estimates and speeds up convergence.

Other potential biomedical applications such as the analysis of olfactory data and magnetoencephalographic (MEG) data are subject of future investigations.

Notes

1. MRI is a method to obtain images of various parts of the body without the use of x-rays. In contrast to x-rays and CAT scans, a MRI scanner consists of a large and very strong magnet in which the patient lies. A radio wave antenna is used to send signals to the body and then receive signals back. Given the received signals which are changing magnetic fields that are much weaker than the steady strong magnetic field of the main magnet, an image of the body can be computed at almost any particular angle.

2. Functional MRI refers to the use of MRI scans with a specific task such as psychomotor tasks to observe brain activity during performance. fMRI detects subtle increases in blood flow associated with activation of parts of the brain and may be useful for preoperative neurosurgical planning, epilepsy evaluation, and "mapping" of the brain.

3. The basis of the BOLD technique lies on the fact that MRI images can be made sensitive to local oxygen concentrations in tissue. This effect has been applied almost exclusively in the human brain to map cortical regions responsible for performing various cognitive tasks, since the oxygenation level in active cortex changes between baseline and tasking conditions.

4. Since the task is an alternating visual task the reference function is expected to exhibit a square wave type signal

5. One may ask why ICA results in correlated data. The answer is that a spatial ICA was performed, assuming independent image maps in contrast to temporal ICA giving temporally independent components.

8 ICA FOR FEATURE EXTRACTION

On ne voit bien qu'avec le coeur.
L'essentiel est invisible pour les yeux.
Antoine de Saint-Exupéry

8.1 OVERVIEW

Barlow (1961) proposed that the goal of sensory coding is to transform the input signals such that it reduces the redundancy between the inputs. Atick (1992) and Atick and Redlich (1993) have used correlation-based methods suggesting that the principle of redundancy reduction may be applied towards the understanding of coding principles in retinal cells in the visual cortex (Field, 1994). This strategy may be used for the purpose of efficient coding for natural images since they are not purely random but contain structure. Natural images have oriented lines, edges and other structures that have dependencies of higher-order statistics. These localized structures can be described mainly by the phase spectrum. For example, an edge occurs locally and has its phase spectrum aligned across different spatial frequencies. Correlation-based approaches are phase-blind, i.e. they can capture only the power spectrum and higher-order methods are needed to account for localized oriented structures. Olshausen and Field (1996) considered a network that maximizes the sparseness of the representation of natural images and showed that the extracted features are localized and oriented. Those features are similar to receptive fields in the primate striate cortex.

Along this line of research Bell and Sejnowski (1997) suggest that *independent components of natural scenes are edge filters.* They apply the infomax learning rule (Bell and Sejnowski, 1995) to an ensemble of natural scenes and find sets of visual filters (features) that are localized and oriented. Compared to results from decorrelation-based methods such as PCA, the ICA filters are sparsely distributed outputs and similar to Gabor filters [1] found by Olshausen and Field (1996) using the sparseness-maximization network.

This chapter reviews the suggestion of Bell and Sejnowski (1997) that ICA can be used to extract features (filters) from natural scenes that are localized edge detectors. Their results are briefly summarized and confirmed by using the extended infomax algorithm (eq.2.56) that the independent component of natural images are sparsely localized filters. To this end, both algorithms are applied to 10 natural images and compare their features. Similar findings have been reported by Hyvaerinen and Oja (1997b) and Karhunen et al. (1997b) using the fixed-point algorithm . This is not surprising since their algorithm can as well separate sub- and super-Gaussian components. The fixed-point algorithm (Hyvaerinen and Oja, 1997a) is closely related to Comon (1994) cumulant maximizing approach discussed in chapter 3.

Bartlett and Sejnowski (1997) and Gray et al. (1998) demonstrate the successful use of ICA filters as features in face recognition tasks and lipreading tasks respectively.

The organization of the chapter is as follows: The methods suggested by Bell and Sejnowski (1997) are reviewed in section 8.2. The extended infomax algorithm is applied to 10 natural images in section 8.3. A brief summary is given for the use of ICA features in face recognition and lip-reading tasks in section 8.4. Other potential applications are discussed in section 8.5.

8.2 ICA OF NATURAL IMAGES

The methods and obtained results by Bell and Sejnowski (1997) are briefly reviewed. Their proposed method extract features from natural scenes by assuming a linear image synthesis model (Olshausen and Field, 1996). Such a model is shown in figure 8.2 where each patch $\mathbf{x}$ of an image is a linear combination of several underlying basis functions. Here, the columns of $\mathbf{A}$ are the basis functions and the image patch is generated according to the following linear model

$$\mathbf{x} = \mathbf{A}\mathbf{s} \tag{8.1}$$

where $\mathbf{s}$ represents the weightening of the basis functions $\mathbf{A}$ to form $\mathbf{x}$. $\mathbf{s}$ can be thought as 'causes' of the images where a linear synthesis of these causes constitute the observations $\mathbf{x}$. Figure 8.1 shows the linear image synthesis model with the columns of $\mathbf{s}$ representing the causes, rows of $\mathbf{A}$ are the basis functions and the columns of $\mathbf{x}$ are the observed image patches. The goal of the recognition model in figure 8.3 is to find a matrix $\mathbf{W}$ so that $\mathbf{u} = \mathbf{W}\mathbf{x}$ are the underlying causes when $\mathbf{W} = \mathbf{A}^{-1}$. The basis functions are then rows of $\mathbf{W}$. Therefore, learning $\mathbf{W}$ gives

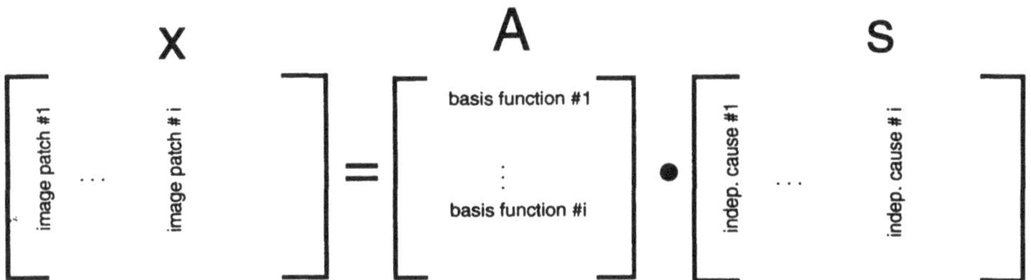

Figure 8.1. Linear image synthesis model with causes $\mathbf{s}$, basis functions $\mathbf{A}$ and observed image patches $\mathbf{u}$.

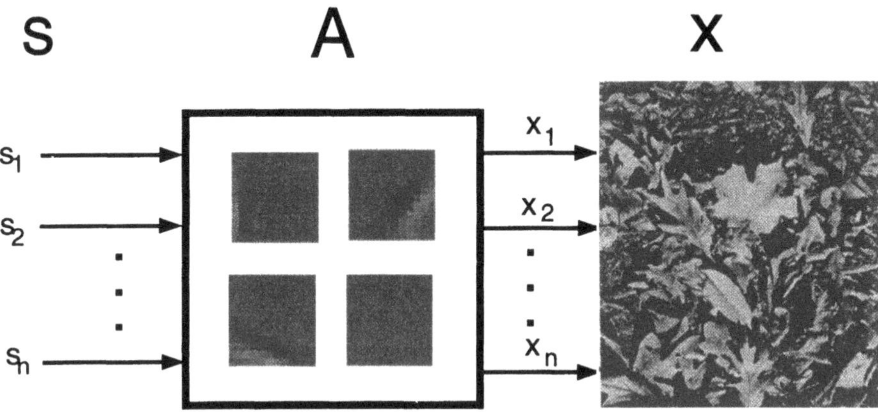

Figure 8.2. Cartoon of a linear synthesis model. Images are composed linearly of basis functions $\mathbf{A}$ mixed with the amplitudes of the causes s_i. For simplicity, only 4 basis functions are shown.

rise about what constitutes a 'cause'. In terms of the visual processing strategy the rows of $\mathbf{W}$ are the receptive fields.

Bell and Sejnowski (1997) applied the infomax learning rule in eq.2.46 to four natural images [2]. Their methodology and their main results are well described in Bell and Sejnowski (1997).

8.3 ICA OF NATURAL IMAGES USING EXTENDED INFOMAX

The goal in this section is to repeat the experiment in Bell and Sejnowski (1997) using the extended infomax algorithm in eq.2.56 (Lee et al., 1998b). The motivation is to find in addition to sparse super-Gaussian representations also low-kurtotic sub-Gaussian representations.

Ten natural images including the four images used in Bell and Sejnowski (1997) are taken for this experiment. 16.700 patches of size 12×12 pixels from those images were randomly selected. Each sample is a column in the observation matrix

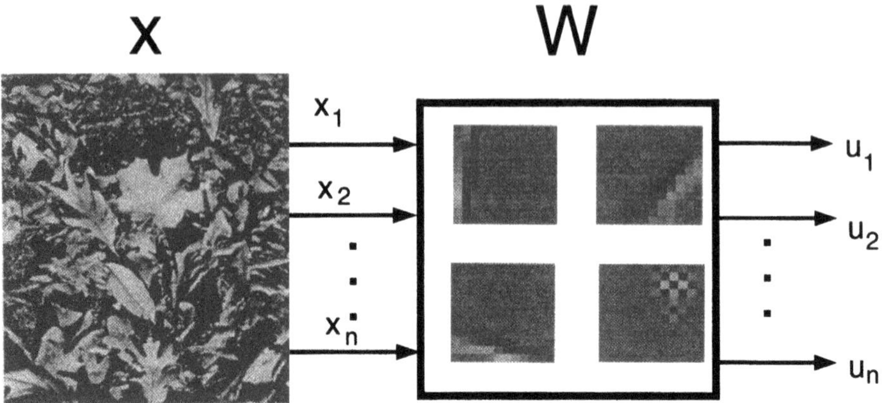

Figure 8.3. Cartoon of an image recognition model. The goal of efficient coding is to learn the basis functions $\mathbf{W}$ to recover the 'causes' s_i.

$\mathbf{x}$ (144×16700). The basis functions $\mathbf{W}$ was learned using the extended infomax algorithm in eq.2.56 with the following parameters: block size of 50, a fixed learning rate of 0.0001 and 100 passes through the data (33400 weight updates). A momentum term was used as well. The learned $\mathbf{W}$ had a dimension of 144×144. Figure 8.4 shows the complete set of learned filters.

Each filter represents a row of $\mathbf{W}$ and they are ordered in length of the filter vector where the filters at the top correspond to filters producing a higher entropy at the outputs. The ICA filters are localized and mostly oriented. In this order, the filters consist of one DC filter (top left), 3 filters close to the DC filter (2-4) and 72 oriented filters. 46 of them were diagonal, 12 are vertical and 14 horizontal. There are also 48 localized checkerboard patterns. The outputs of the filters $\mathbf{u}$ are mostly sparse distributed. The output of the DC component was close to a Gaussian distribution and all the other outputs signals were kurtotic. The extended infomax algorithm detected the DC component as one sub-Gaussian component. The kurtosis values ranged from low kurtotic ($k_4 = -0.06$ for the output of the DC filter) up to high kurtotic outputs ($k_4 = 15$) for the output of the checkerboard-like filters. The mean kurtosis for all u_i's was 8.8. Roughly the same results have been reproduced by using the original infomax learning rule. The DC filter learned by using the original infomax algorithm (Bell and Sejnowski, 1995) was very similar to the low kurtotic output detected by the extended infomax algorithm. These results suggest that independent components in natural images are almost solely generated by sparse filters.

8.4 ICA FOR FEATURE EXTRACTION

A valid question one may ask is if the learned filters (features) $\mathbf{W}$ can be used as an efficient code to discriminate the observed data. In other words, can the outputs

Figure 8.4. ICA on natural images using extended infomax. The matrix of 144 filters obtained by training on 16.700 random samples from natural images of size 12×12 pixels. Each filter is a row of the matrix $\mathbf{W}$, and they are ordered left-to-right, top-to-bottom in reverse order of the length of the filter vector.

of an ICA transformation be used to discriminate observations? Furthermore, is it possible to use the ICA features as information in classification systems so that more reliable decisions are made?

Bartlett and Sejnowski (1997) and Gray et al. (1998) demonstrate the successful use of ICA filters as features in face recognition tasks and lipreading tasks respectively. A brief summary of their methods follows:

- **Example Face Recognition**

The goal in face recognition is to train a system so that it can recognize and classify familiar faces given a different image of the trained face. The test images may show faces in a different pose or under different lighting conditions. Traditional methods to face recognition have employed PCA-like methods Turk and Pentland (1991). Bartlett and Sejnowski (1997) compare the face recognition performance of PCA and ICA for two different tasks: (1) different pose and (2) different lighting condition. They show that for both tasks ICA outperforms PCA. The method is roughly as follows: The rows of the face images constitute the data matrix $\mathbf{x}$. Performing ICA, a transformation $\mathbf{W}$ is learned so that $\mathbf{u}$ ($\mathbf{u} = \mathbf{W}\mathbf{x}$) represent independent face images. Nearest neighbor classification is performed on the coefficients of $\mathbf{u}$. In comparing PCA and ICA, nearest-neighbor classification of the ICA representation outperformed the PCA representation.

- **Example Lipreading**

The goal in the lipreading task is to recognize a word given only a sequence of lip images that are observable when the word is spoken. The recognizer is first trained on features of the lip images labeled according to the spoken word. Then, the performance of the recognizer is tested by classifying a new sequence of lip images. Gray et al. (1998) have compared local and global PCA and ICA representations. Their method can be roughly summarized as follows: They first extract features that represent the images: Optical flow representations have been used to account for temporal processing. Spatial information have been extracted using either local or global PCA or ICA. In global ICA/PCA one can think of an ensemble of entire lip images $\mathbf{x}$ that is decomposed into independent components $\mathbf{u}$ using a linear transformation $\mathbf{W}$ where the rows of $\mathbf{W}$ are global features. Local ICA/PCA involves small patches of the lip images to extract $\mathbf{W}$. Only one of four classes (global ICA/PCA and local ICA/PCA)) of features are then used to train the parameters in an Hidden Markov Models (HMM). HMM are probabilistic methods that enable to model the observed features in probabilistic states and probabilistic transitions between states. A certain parametric combination of states and transitions determine the probability of the observation of certain features that are associated with features from a labeled training example. Their result suggest that local features are more suitable for their task and that local PCA performed slightly better than local ICA.

8.5 DISCUSSIONS AND CONCLUSIONS

Results were presented using the extended infomax algorithm that corroborate the findings of Bell and Sejnowski (1997) and Olshausen and Field (1996) that the independent components of natural scenes are localized and oriented edge detectors.

Bell and Sejnowski (1997) included many discussion issues and additionally a detailed comparison with decorrelation-based methods (PCA and zero-phase whitening filters). An approach proposed by Lewicki and Sejnowski (1998b) is able to learn an overcomplete bases to find efficient sound codes as well as image codes in audio signals and images respectively (chapter 5). Overcomplete bases have a greater number of basis vectors than the dimensionality of their input vectors Simoncelli et al. (1992). For natural images Lewicki and Olshausen (1998) show that the overcomplete bases are localized and oriented and confirm the results obtained earlier by Olshausen and Field (1996) and Bell and Sejnowski (1997) using complete bases. In contrast to complete bases they show that higher degrees of overcompleteness can better model the observed data. An application to noise removal is presented in Lewicki and Olshausen (1998) where they demonstrate that the learned bases form efficient codes that have better denoising properties than traditional complete basis and overcomplete Fourier and wavelet bases. Potential applications of this technique include new data compression techniques.

There are several issues subject to future research efforts. Edges are the first level of invariance and can be detected by a linear transformation. However, one might wish to find transformation that are invariant to shifting, scaling and rotation (Bell and Sejnowski, 1997). From a biology point of view the discussed algorithms are non-local, i.e. the neuron makes use of information which is present in other neurons without having the necessary connection. Therefore, the algorithms have to be reconsidered for image processing issues in the visual cortex.

In a similar manner, Bell and Sejnowski (1996) applied the infomax algorithm to learning higher-order structure of a natural sound.

Hateren and Ruderman (1998) performed ICA on time varying images (video) and showed that ICA yields spatiotemporal filtering like in simple cells.

Notes

1. Gabor wavelets are of similar shape as the receptive fields of simple cells in the primary visual cortex (V1). They are localized in both space and frequency domains and have the shape of plane waves restricted by a Gaussian envelope function.

2. The images are available in (`ftp://ftp.cnl.salk.edu/pub/tony/VRimages`).

9 UNSUPERVISED CLASSIFICATION WITH ICA MIXTURE MODELS

Zwei Seelen wohnen; ach; in meiner Brust.
Goethe. "Faust"

9.1 OVERVIEW

This chapter presents an unsupervised classification algorithm based on an ICA mixture model (Lee et al., 1998d). The mixture model is a model in which the observed data can be categorized into several data classes. The data in each class is generated by a linear mixture of independent sources and is called an ICA class. The goal of the learning algorithm is first to find the independent sources and the mixing matrix for each ICA class and second to compute the posterior probability of the class membership of a given data point. The first step of the algorithm employs a standard ICA algorithm to learn the basis vectors (mixing matrix) and the bias terms. In the second step, the algorithm computes for each ICA class the probability of the ICA class given the data. This approach can be seen as an extension of the Gaussian mixture model in which the clusters can have non-Gaussian structure. Performance on a standard classification problem, the Iris flower data set, demonstrate that the new algorithm achieves highly competitive classification results.

ICA is a technique to find a linear non-orthogonal co-ordinate system in multivariate data. The directions of the axes of this co-ordinate system are determined by both the second and higher order statistics of the observed data $\mathbf{x}$. The goal is to linearly transform the data such that the resulting data $\mathbf{s}$ is as statistically independent from each other as possible (Bell and Sejnowski, 1995; Cardoso and Laheld, 1996; Lee et al., 1998a).

Successful applications of separating mixed speech signals (Lee et al., 1997a) and removing artifacts from EEG recordings (Jung et al., 1998a) have demonstrated that ICA is a useful tool to find independent components in real world data. Girolami (1997b) showed how ICA can be used to find independent clusters in data sets. Here, a new learning algorithm is presented that can be used to identify several classes in the observed data in such a way that the data in each class are generated by a mixture of independent components.

9.2 THE ICA MIXTURE MODEL

Assume that there are K classes of M-dimensional zero-mean vectors $\mathbf{s}_k$, such that the components s_i within each class are mutually independent. Each class of independent sources $\mathbf{s}_k$ corresponds to M independent scalar-valued sources s_i. The multivariate p.d.f. of the vector can be written as the product of marginal independent distributions $p(\mathbf{s}|\Omega) = \prod_{i=1}^{M} p(s_i|\omega_i)$ where $\Omega = \omega_1, \cdots, \omega_M$ is the set of parameters describing the density of $\mathbf{s}$.

$$\begin{aligned} \mathbf{s}_k &= [s_1, \cdots, s_m, \cdots, s_M]^T, m = 1, \cdots, M \\ \mathbf{S} &= [\mathbf{s}_1, \cdots, \mathbf{s}_k, \cdots, \mathbf{s}_K]^T, k = 1, \cdots, K. \end{aligned}$$

Define an ICA class $\mathbf{x}_k$ (N-dimensional vector) as a class in which the independent components $\mathbf{s}_k$ are mixed linearly by a mixing matrix (basis vectors).

$$\mathbf{x}_k = \mathbf{A}_k \mathbf{s}_k + \mathbf{b}_k, \tag{9.1}$$

where $\mathbf{A}_k$ is a full rank $N \times M$ scalar matrix and $\mathbf{b}_k = [b_1, \cdots, b_N]$ is the bias term. An ICA mixture model is defined as follows

$$\mathbf{X} = [\mathbf{x}_1, \cdots, \mathbf{x}_k, \cdots, \mathbf{x}_K]^T, k = 1 \cdots K, \tag{9.2}$$

where $\mathbf{x}_k$ is an ICA class with

$$\mathbf{x}_k = [x_1, \cdots, x_n, \cdots, x_N]^T, n = 1 \cdots N. \tag{9.3}$$

For simplicity, consider here the case where the number of sources is equal to the number of mixtures $N = M$. Here, $\mathbf{X}$ denotes all observed data, i.e. K sets of observations $\mathbf{x}_k$ each being generated by classes of mixed independent sources $\mathbf{s}_k$. The mixing matrix $\mathbf{A}_k$ and the bias term $\mathbf{b}_k$ is different for each class.

Figure 9.1 shows a simple example of classifying an ICA mixture model. There are 2 ICA classes (+) and (o), where each class was generated by two independent variables, 2 bias terms and 2 basis vectors. Class (o) was generated by 2 uniform distributed sources whereas class (+) was generated by 2 Laplacian distributed sources.

In order to correctly classify the observed data $\mathbf{X}$ the task is to learn the following

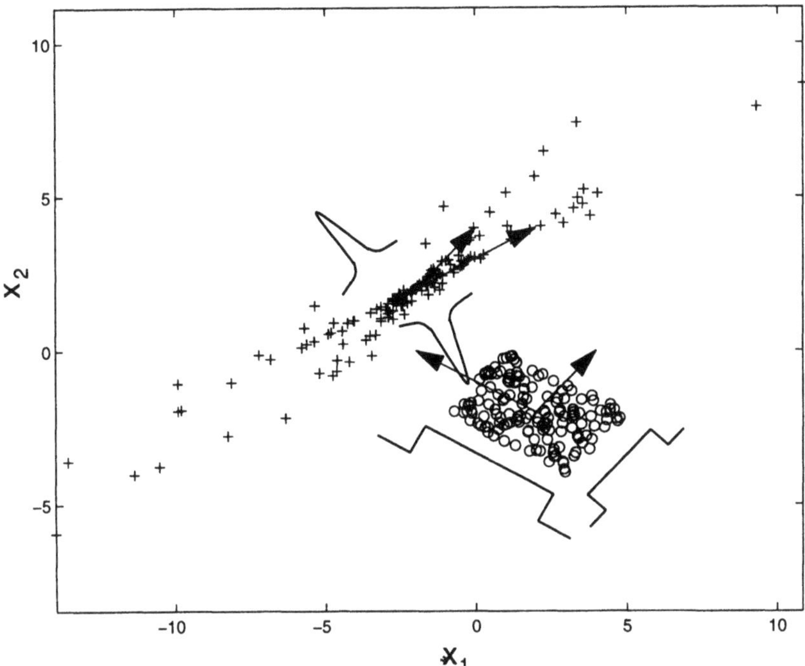

Figure 9.1. A simple example for classifying an ICA mixture model. There are 2 ICA classes (+) and (o), each class was generated by two independent variables, 2 bias terms and 2 basis vectors. Class (o) was generated by 2 uniform distributed sources as indicated next to the data class. Class (+) was generated by 2 Laplacian distributed sources with a sharp peak at the bias and heavy tails.

1. The ICA parameters for each class $C_k = \{\mathbf{A}_k, p(\mathbf{s}_k)\}$.

2. The probability of each class $P(C_k|\mathbf{x}, \mathbf{A}_k)$ for the given data point.

(1) involves estimating the prior of the sources $\hat{\mathbf{p}}_k(\mathbf{s})$ and learning the mixing matrices $\mathbf{A}_k$ using a standard ICA algorithm. Furthermore, the bias terms $\mathbf{b}_k$ must be learned adaptively as well. (2) involves learning the likelihood of the data sample for each class given the model estimates from (1).

For example, assume that the basis functions $\mathbf{A}_k$ and the bias terms $\mathbf{b}_k$ are learned correctly as indicated by the arrows in figure 9.1. If the test data sample is drawn from the uniform distribution (o) then the log-likelihood of class (o) is greater than the log-likelihood of class (+) ($\log P(\mathbf{x}|C_o, \mathbf{A}_o,) > \log P(\mathbf{x}|C_+, \mathbf{A}_+,)$). The posterior probability of each class (C_o, C_+) can be computed given the appropriate log-likelihoods and the class priors.

The learning algorithm can be derived by an expectation maximization (EM) approach Ghahramani (1994) and implemented in the following steps:

1. Compute the log-likelihood of the data for each class

$$\log P(\mathbf{x}|C_k, \mathbf{A}_k) = \log P(\mathbf{s}) - \log(\det|\mathbf{A}|) \tag{9.4}$$

2. Compute the probability for each class

$$\log P(C_k|\mathbf{x}, \mathbf{A}_k) = \frac{P(\mathbf{x}|\mathbf{A}_k, C_k)P(C_k)}{\sum_k P(\mathbf{x}|\mathbf{A}_k, C_k)P(C_k)} \tag{9.5}$$

3. Learn adaptively the basis functions $\mathbf{A}$ and the bias terms $\mathbf{b}$ for each class. The basis functions $\mathbf{A}$ are learned according to the extended infomax ICA learning rule by Lee et al. (1998b) and by taking into account the probability for each class

$$\begin{aligned} \mathbf{A}_k &\approx \frac{\partial}{\partial \mathbf{A}} P(\mathbf{x}|\mathbf{A}_k) && (9.6) \\ &= P(C_k|\mathbf{x}, \mathbf{A}_k)\frac{\partial}{\partial \mathbf{A}} P(\mathbf{x}|C_k, \mathbf{A}_k). && (9.7) \end{aligned}$$

   Note that any ICA learning algorithm can be used to learn the gradient. For the bias term the learning rule is

$$\mathbf{b}_k = \frac{\sum_t x_t P(C_k|x_t)}{\sum_t P(C_k|x_t)}, \tag{9.8}$$

   where t is the index in the training data $(1, \cdots, t, \cdots, T)$.

The three steps in the learning algorithm increase the total likelihood of the data.

$$P(\mathbf{x}|\mathbf{A}_{1:K}, C_{1:K}) = \sum_k P(\mathbf{x}|A_k, C_k)P(C_k). \tag{9.9}$$

The extended infomax ICA learning rule (see chapter 2) is able to blindly separate mixed sources with sub- and super-Gaussian distributions. This is achieved by using a simple type of learning rule first derived by Girolami (1998). The learning rule in Lee et al. (1998b) uses the stability analysis of Cardoso and Laheld (1996) to switch between sub- and super-Gaussian regimes. The learning rule for $\mathbf{W}$ (an estimate of the inverse of $\mathbf{A}$) is

$$\Delta\mathbf{W} \propto \left[\mathbf{I} - \mathbf{K}\tanh(\mathbf{u})\mathbf{u}^T - \mathbf{u}\mathbf{u}^T\right]\mathbf{W} \left\{ \begin{array}{rcl} k_i = 1 & : & \text{super} - \text{Gaussian} \\ k_i = -1 & : & \text{sub} - \text{Gaussian} \end{array} \right. \tag{9.10}$$

where k_i are elements of the N-dimensional diagonal matrix $\mathbf{K}$ and $\mathbf{u} = \mathbf{W}\mathbf{x}$. The k_i's are (Lee et al., 1998b)

$$k_i = \text{sign}\left(E\{\text{sech}^2(u_i)\}E\{u_i^2\} - E\{[\tanh(u_i)]u_i\}\right). \tag{9.11}$$

For the log-likelihood estimation in eq.9.4 the term $P(\mathbf{s})$ can be approximated as follows

$$\begin{array}{rcl} \log P(\mathbf{s}) \approx -\sum_n |s_n| & : & \text{super} - \text{Gaussian} \\ \log P(\mathbf{s}) \approx -\sum_n \left(\log(\cosh(s_n)) - \frac{s_n^2}{2}\right) & : & \text{sub} - \text{Gaussian} \end{array} \tag{9.12}$$

For super-Gaussian densities, the approximation is achieved by a Laplacian density model. In case of sub-Gaussian densities, $P(\mathbf{s})$ is approximated by a bimodal density of the form in eq.2.57. Although the source density approximation is crude it has been demonstrated that simple density models are sufficient for standard ICA problems (Lee et al., 1998b).

The testing procedure this accomplished by processing each the test data sample using learned parameters $\mathbf{A}_k$ and $\mathbf{b}_k$. The probability of the class $P(C_k|\mathbf{x}, \mathbf{A}_k)$ is computed and the corresponding label is compared to the highest class probability.

9.3 SIMULATIONS

To demonstrate the validity of the learning algorithm random data was generated and drawn from different classes. The steps in the above section were used to learn the parameters and to classify the data. Figure 9.2 gives a simulation example of how several classes may be represented in a two-dimensional data space. Each class was generated as two random variables with an arbitrary density function. Then, for each class the two variables were mixed by a random mixing matrix $\mathbf{A}$ and a random bias vector $\mathbf{b}$. In this example, the algorithm had to find four mixing matrices and four bias vectors given only the two dimensional data set. To verify the classification process, the data was divided into training and testing data sets. For training, the parameters were randomly initialized. Given the observed data set $\mathbf{X} = [\mathbf{x}_1, \cdots, \mathbf{x}_4]$ the algorithm converged in 30 iterations through the data. The arrows in figure 9.2 indicate the basis vectors $\mathbf{A}_k$ and the bias terms $\mathbf{b}_k$ were found correctly for the different classes. Testing was accomplished by processing each instance with the learned parameters $\mathbf{A}_k$ and $\mathbf{b}_k$. The probability of the class $P(C_k|\mathbf{x}, \mathbf{A}_k)$ was computed and the corresponding instance label was compared to the highest class probability. For this simulation example the classification error on the test set was 7.5% although the classes had several overlapping areas. For comparison with other methods, the same data were applied using the k-means (Euclidean distance measure) clustering algorithm optimized with the EM algorithm. This method gave an error of 11.3%.

9.4 IRIS DATA CLASSIFICATION

To compare the proposed method to other classification algorithms, the method has been applied to classify real data from the machine learning benchmark (Merz and Murphy, 1998). As an example the classification of the well known iris flower data set is shown here. The data set (Fisher, 1936) contains 3 classes, 4 numeric attributes of 50 instances each, where each class refers to a type of iris plant. One class is linearly separable from the other two, but the other two are not linearly separable

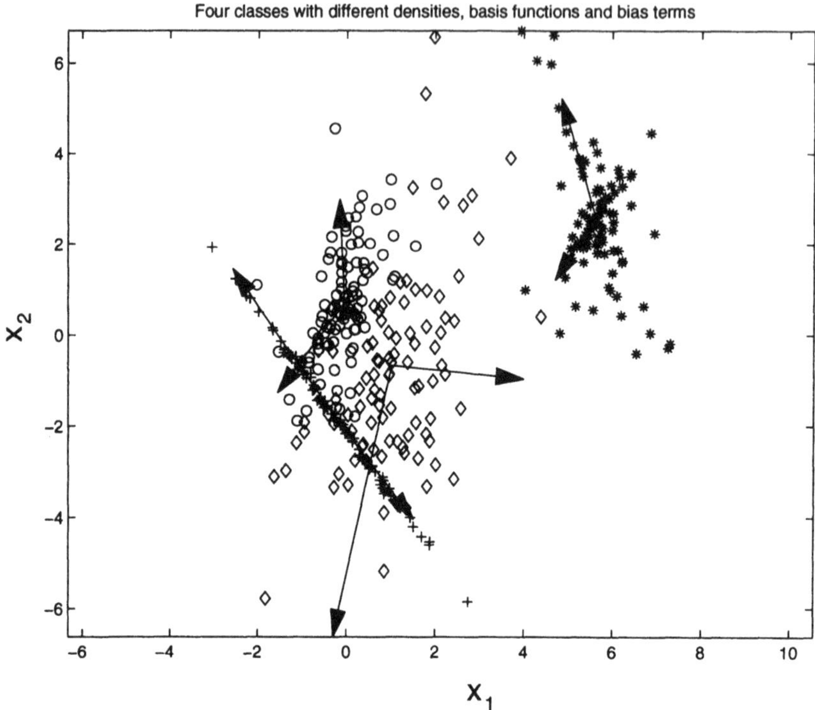

Figure 9.2. An example of classification of a mixture of independent components. There are 4 different classes, each generated by two independent variables and bias terms. The algorithm is able to find the independent directions (basis vectors) and bias terms for each class.

from each other. The complete data set was divided into training set (75%) and testing set (25%). The algorithm converged after one hundred passes through the training data. The classification error on the training data set was 2% whereas the error on the test data set was 3.3%. For comparison, Freund (1995) used a simple classifier with an additional boosting method and reported an error rate of 4.8%. A k-means clustering using the EM algorithm gave an error rate of 4.12%.

This is a classical data set in statistics and the improvement in performance from the previous reported results is highly significant.

9.5 CONCLUSIONS

The new algorithm for unsupervised classification presented here is based on a maximum likelihood mixture model using independent component analysis to model the structure of the clusters. The algorithm demonstrated on simulated and real world data that its classification results are highly competitive.

This new method is similar to to other approaches such as the mixture density networks by Bishop (1994) in which a neural network was used to find arbitrary density functions. The two steps of the proposed algorithm can be derived from the EM algorithm for density estimation (Ghahramani, 1994; Amari, 1995).

This algorithm reduces to the Gaussian mixture model when the source priors are Gaussian. Purely Gaussian structure, however, is rare in real data sets. Here priors of the form of super-Gaussian and sub-Gaussian densities were used. But these could be extended as proposed by Moulines et al. (1997) and Attias (1998). The proposed model was used for learning a complete set of basis functions without additive noise. However, the method can be extended to take into account additive Gaussian noise and an overcomplete set of basis vectors (Lewicki and Sejnowski, 1998b).

Other applications of the proposed method include modeling the context switching in blind source separation which occurs when the observed signals are mixed with a non-stationary mixing matrix. This can be significant in the automatic detection of sleep stages by observing EEG signals. The ICA mixture model may be used to automatically identify these stages due to the changing source priors and their mixing.

10 CONCLUSIONS AND FUTURE RESEARCH

In the end is my beginning
T.S. Eliot

10.1 CONCLUSIONS

Theories and applications of ICA were presented. The first part of the book focused on unsupervised learning algorithms for ICA. Based on fundamental theories in probabilistic models, information theory and artificial neural networks several unsupervised learning algorithms such as infomax, maximum likelihood estimation, negentropy maximization, nonlinear PCA, Bussgang algorithm and cumulant-based methods are presented that can perform ICA. Those seemingly different theories are reviewed and put in an information theoretic framework to unify several lines of research. An extension of the infomax algorithm of Bell and Sejnowski (1995) is presented that is able to blindly separate mixed signals with sub- and super-Gaussian source distributions (Girolami, 1997b; Lee et al., 1998b). The learning algorithms are furthermore extended to deal with the multichannel blind deconvolution problem. The use of filters allows the separation of voices recorded in a real environment (cocktail party problem). Although the ICA formulation has several constraints such as the linear model assumption, the number of sensors and the low-noise assumption, it can be demonstrated that new methods can loosen some constraints. In particular, an overcomplete representation of the ICA formulation (Lewicki and Sejnowski,

1998c) can be used to represent more basis functions than the dimensionality of the data. This method is therefore able to model and to infer more sources than sensors. The advantage of the inference model is that it includes a noisy ICA model and is therefore able to cope with additive noise. However, the overcomplete representation appears more sensitive to the source density mismatch than the complete ICA representation. A few steps towards nonlinear ICA were presented. This issue is ill conditioned and has in general not a unique solution. However, given appropriate constraints there are certain solvable models such as the two stage model Lee et al. (1997c) in which a nonlinear transfer function follows after the linear mixing.

The second part of the book presented applications of ICA to real-world problems. The ICA algorithm has been successfully applied to many biomedical signal processing problems such as the analysis of electroencephalographic (EEG) data (Makeig et al., 1997; Jung et al., 1998a). Makeig et al. (1996) have applied the original infomax algorithm (Bell and Sejnowski, 1995) to EEG and ERP data showing that the algorithm can extract EEG activations and isolate artifacts. Jung et al. (1998a) show that the extended infomax algorithm (Lee et al., 1998b) is able to linearly decompose EEG artifacts such as line noise, eye blinks, and cardiac noise into independent components with sub- and super-Gaussian distributions. McKeown et al. (1998b) have used the extended ICA algorithm to investigate task-related human brain activity in fMRI data. Another area of applications can result from exploring independent features in images. Bell and Sejnowski (1997) suggested that independent components of natural scenes are edge filters. Those features may be used in pattern classification problems such as visual lip-reading and face recognition tasks to improve its recognition performance. The ICA algorithm can be furthermore embedded in an expectation maximization framework with the goal to classify clusters of ICA models. This approach is an extension of the Gaussian mixture model for non-Gaussian priors. Results on classification benchmarks demonstrate that ICA cluster models can improve classification results.

Although several limitations and assumptions impedes the use of ICA, it seems appropriate to conjecture that the algorithms and methods are useful tools with many potential applications where many second-order statistical methods reach their limits. Potential applications are now being optimized and may take a few more years until they will be commercially used. Several researchers believe that these techniques will have a huge impact on engineering methods and industrial applications.

10.2 FUTURE RESEARCH

There are many issues subject to further investigation. As pointed out in chapter 2 ICA relies on several model assumptions that may be inaccurate or even incorrect. Those issues are highlighted and their potential solutions are discussed.

- **Nonlinear mixing problem**

  Researchers have recently tackled the problem of nonlinear mixing phenomena. Burel (1992); Lee et al. (1997c); Taleb and Jutten (1997); Yang et al. (1997); Hochreiter and Schmidhuber (1998) propose extensions when linear mixing is combined with certain nonlinear mixing models. Other approaches include self-organizing feature maps to identify nonlinear features in the data (Hermann and Yang, 1996; Lin and Cowan, 1997; Pajunen, 1996). Chapter 6 presented some simplified models and their solutions. However, the methods are far from being generally applicable. The nonlinear ICA problem needs to be well defined and constrained to a solvable solution. Furthermore, the independence assumption and the nonlinear ICA model are two contradictory terms because there are no more unique solutions to nonlinear ICA models, e.g. the nonlinearly mixed signals may be linearly independent. Perhaps drastically different principles have to be addressed so that sources can be separated by self-organizing principles that are not necessarily relying on information-theoretic principles.

- **Underdetermined ICA**

  The underdetermined problem in ICA, i.e. having more sources than sensors $N < M$ is of theoretical and practical interest. The overcomplete ICA representation (Lewicki and Sejnowski, 1998c) is highly promising. However, its implementation is rather complicated and other methods may be required that use a priori knowledge of the source distribution. In particular, semi-blind methods, i.e. method that use a priori information about the source density or the temporal structure of the source, are of interest. The incorporation of temporal structure is key to solving the single channel source separation problem. An approach by Lewicki and Sejnowski (1998a) indicate first steps towards solving this inherently difficult problem.

- **Noisy ICA**

  Only a few papers have discussed ICA in the presence of additive noise (Nadal and Parga, 1994; Attias, 1998) and much more work needs to be done to determine the effect of noise on performance. Although the overcomplete ICA framework by Lewicki and Sejnowski (1998c) uses an additive noise model the inference is based on the sparse source assumption it needs to be extended for general source separation issues. There may be other generative models Hinton and Ghahramani (1997) in a Bayesian framework that can cope with additive noise. A very promising idea to solve this problem is the independent factor analysis by Attias (1998). It is a generalization of factor analysis, PCA and ICA in which the model parameters are learned using an EM algorithm.

- **Non-stationarity problem**

  Sources may not be stationary, i.e. sources may appear, disappear or move (speaker moving in a room). In these cases, the weight matrix $\mathbf{W}$ may change completely from one time point to the next. Unsupervised methods are required that take into account abrupt changes in real environments. Nadal and Parga (1998) have proposed some analytical method for time-dependent mixtures. Murata et al. (1997) suggest an adaptation of the learning rate to cope with changing environments. Matsuoka et al. (1995) use a neural network to separate nonstationary signals. A promising off-line method to analyze the non-stationarity problem in ICA is the ICA mixture model presented in chapter 9. The different classes may account for the different mixing models or changing mixing environments.

There are several potential applications that have not been investigated extensively yet but may be of great significance. It is conjectured that ICA can be applied to find independent signals in any multi-sensory array recordings of real data. Furthermore, ICA may be useful to find structure in high dimensional data space for data mining purposes, e.g. structures in medical databases that may be significant for clinical patient evaluation. A few applications for further investigations are suggested.

- **ICA for spike train separation**

  Brown et al. (1988) used ICA to separate action potential trains in multiple-neuron, multiple-detector recordings. They showed that the extended infomax ICA algorithm can separate the recordings into single neurons. Similar results were obtained by Laubach and Nicolelis (1998) where they applied ICA to multi-sensory spike trains recorded in the motor cortex of rats. Their finding suggests that neuronal interactions are distributed sparsely within the motor cortex.

- **ICA on recordings from the olfactory system**

  Hopfield (1991) suggested that the olfactory computation may be related to factorial code representation. The application of ICA to data from the olfactory system is currently investigated (Kauer and White, personal communication) to test the hypothesis.

- **ICA in communications**

  Complex valued signal mixing occurs in radio channels. This is a problem in current mobile communication applications such as CDMA (Code Division Multiple Access) systems. Torkkola (1998) incorporated prior knowledge about the source distributions into the nonlinear transfer function and adaptively found time-varying mixing matrices. In simulations, he showed that infomax can be successfully applied to unmix radio signals in fading channels.

- **ICA for data mining**

  Data mining, the extraction of hidden predictive information from large databases, is a powerful new technology with great potential to help companies focus on the most important information in their data warehouses. Lizhong and Moody (1997) explored ICA for financial data modeling. More recently, Girolami (1997a) suggested projection pursuit networks for data mining. They compared their data cluster projections by ICA and found different results which were more significant than other traditional methods. Isbell and Viola (1998) used ICA for text retrieval in high-dimensional text data bases. The goal is to find a subset of a collection of documents relevant to a user's information request. This technique may have an impact on Internet search engines. The ICA mixture model in Lee et al. (1998d) can be applied to general data mining purposes. Regarding the findings of the simple ICA methods it is assumed that the ICA mixture model will be able to extend the formulation into more unsupervised classification problems.

- **Biological evidence of ICA?**

An interesting question from a neuroscience viewpoint is the understanding of learning mechanisms with factorial codes. Although the learning rules in eq.2.46 and eq.2.56 in a single feedforward architecture are non-local, i.e. the neurons must know information about the synaptic weights of neighboring neurons without being connected to them. There are a few local learning rules for ICA. The Herault-Jutten architecture has a local learning rule. The extended exploratory projection pursuit network with inhibitory lateral connections (Girolami and Fyfe, 1997b) has a local learning rule as well. Field (1994) suggested that factorial code is an efficient coding strategy for visual sensory processing. There may be further evidence of factorial coding principles in other neurons such as the cerebellum that might use efficient coding schemes for motor control and prediction (Coenen et al., 1998).

- **ICA chip**

A CMOS integration of Herault-Jutten cells was realized by Vittoz and Arreguit (1989). A VLSI chip implementation of the Herault-Jutten algorithm was realized by Cohen and Andreou (1992). However, the stability of the Herault-Jutten algorithm is sensitive with respect to the mixing condition (Sorouchyari, 1991) due to the missing equivariance property. Therefore, current investigations include the implementation of the extended infomax algorithm in VLSI. An extension to time-delays and convolved mixtures may be of practical interest and will be addressed in the near future.

Bibliography

Amari, S. (1995). Information geometry of the em and em algorithms for neural networks. *Neural Networks*, 8:9:1379–1408.

Amari, S. (1997a). Neural learning in structured parameter spaces. In *Advances in Neural Information Processing Systems 9*, pages 127–133. MIT Press.

Amari, S. (1997b). Superefficiency in blind source separation. *IEEE Trans. on Signal Processing*, submitted.

Amari, S. (1998). Natural gradient works efficiently in learning. *Neural Computation*, in press.

Amari, S. and Cardoso, J.-F. (1997). Blind source separation — semiparametric statistical approach. *IEEE Trans. on Signal Processing*, 45(11):2692–2700.

Amari, S., Chen, T.-P., and Cichocki, A. (1997a). Stability analysis of adaptive blind source separation. *Neural Networks*, 10(8):1345–1352.

Amari, S., Cichocki, A., and Yang, H. (1996). A New Learning Algorithm for Blind Signal Separation. In *Advances in Neural Information Processing Systems 8*, pages 757–763.

Amari, S., Douglas, S., Cichocki, A., and Yang, H. (1997b). Multichannel blind deconvolution and equalization using the natural gradient. In *IEEE International Workshop on Wireless Communication*, pages 101–104.

Aris, R. (1962). *Vectors, Tensors and the Basic Equations of Fluid Mechanics.* Dover Publications.

Atick, J. (1992). Could information theory provide an ecological theory of sensory processing? *Network*, 3:213–251.

Atick, J. and Redlich, A. (1993). Convergent algorithm for sensory receptive field development. *Neural Computation*, 5:45–60.

Attias, H. (1998). Blind separation of noisy mixtures: An em algorithm for independent factor analysis. *Neural Computation*, submitted.

Attias, H. and Schreiner, C. E. (1998). Blind Source Separation and Deconvolution: The Dynamic Component Analysis Algorithm. *Neural Computation*, in press.

Back, A. (1994). Blind deconvolution of signals using a complex recurrent network. In *IEEE Workshop on Neural Networks for Signal Processing*, pages 565–574.

Barlow, H. (1961). *Sensory Communication*, volume Rosenblith, W.A. (editor), chapter Possible principles underlying the transformation of sensory messages, pages 217–234. MIT press.

Bartlett, M. and Sejnowski, T. J. (1997). Viewpoint invariant face recognition using independent component analysis and attractor networks. In *Advances in Neural Information Processing Systems 9*, pages 817–823. MIT Press.

Bell, A. J. and Sejnowski, T. J. (1995). An Information-Maximization Approach to Blind Separation and Blind Deconvolution. *Neural Computation*, 7:1129–1159.

Bell, A. J. and Sejnowski, T. J. (1996). Learning the higher-order structure of a natural sound. *Network: Computation in Neural Systems*, 7:261–266.

Bell, A. J. and Sejnowski, T. J. (1997). The 'independent components' of natural scenes are edge filters. *Vision Research*, 37(23):3327–3338.

Bellini, S. (1994). *Blind Deconvolution*, volume Haykin (editor), chapter Bussgang techniques for blind deconvolution and equalization. Prentice Hall.

Belouchrani, A., Abed Merain, K., Cardoso, J., and E., M. (1993). Second-order blind separation of correlated sources. *Proc. Int. Conf. on Digital Sig. Proc.*, pages 346–351.

Belouchrani, A., Meraim, K., Cardoso, J.-F., and Moulines, E. (1997). A blind source separation technique based on second order statistics. *IEEE Trans. on S.P.*, 45(2):434–444.

Berg, P. and Scherg, M. (1991). Dipole models of eye movements and blinks. *Electroencephalog. clin. Neurophysiolog.*, pages 36–44.

Bienenstock, E., Cooper, L., and Munro, P. (1982). Theory for the development of neuron selectivity: Orientation specificity and binocular interaction in visual cortex. *Journal of Neuroscience*, 2:32–48.

Binder, J. (1997). Neuroanatomy of language processing studied with functional mri. *Clinical Neuroscience*, 4(2):87–94.

Bishop, C. (1994). Mixture density networks. *Technical Report*, NCRG/4288.

Bishop, C. (1995). *Neural Networks for Pattern Recognition.* Oxford University Press, Oxford.

Box, G. and Tiao, G. (1992). *Baysian Inference in Statistical Analysis.* John Wiley and Sons.

Brown, G., Satoshi, Y., Luebben, H., and Sejnowski, T. (1988). Separation of optically recorded action potential trains in tritonia by ica. *5th Annual Joint Symposium on Neural Computation*, in press.

Burel, G. (1992). A non-linear neural algorithm. *Neural networks*, 5:937–947.

Cadzow, J. (1996). Blind deconvolution via cumulant extrema. *IEEE Signal Processing Magazine*, 13(3):24–42.

Cardoso, J. (1998a). High-order contrasts for independent component analysis. *Neural Computation*, in press.

Cardoso, J. and Soloumiac, A. (1993). Blind beamforming for non-Gaussian signals. *IEE Proceedings-F*, 140(46):362–370.

Cardoso, J.-F. (1997). Infomax and maximum likelihood for blind source separation. *IEEE Signal Processing Letters*, 4(4):112–114.

Cardoso, J.-F. (1998b). Blind signal processing: a review. *Proceedings of IEEE.* to appear.

Cardoso, J.-F. (1998c). *Unsupervised adaptive filtering*, chapter Entropic contrasts for source separation. S. Haykin (editor) Prentice Hall. to appear.

Cardoso, J.-F. and Comon, P. (1996). Independent component analysis, a survey of some algebraic methods. In *Proc. ISCAS'96*, pages 93–96.

Cardoso, J.-F. and Laheld, B. (1996). Equivariant adaptive source separation. *IEEE Trans. on S.P.*, 45(2):434–444.

Chan, D., Rayner, P., and Godsill, S. (1996). Multi-channel signal separation. In *Proc. ICASSP*, pages 649–652, Atlanta, GA.

Choi, S. and Cichocki, A. (1997). Blind signal deconvolution by spatio - temporal decorrelation and demixing. In *Proc. IEEE Workshop on NNSP*, pages 426–435.

Churchland, P. S. and Sejnowski, T. J. (1992). *The Computational Brain.* MIT Press, Cambridge, MA.

Cichocki, A., Amari, S., and Cao, J. (1997). Neural network models for blind separation of time delayed and convolved signals. *Japanese IEICE Transaction on Fundamentals*, E-82-A(9).

Cichocki, A., Kasprzak, W., and Amari, S. (1995). Multi-Layer Neural Networks with local adaptive learning Rule for Blind Separation of Source Signals. In *Intern. Symp. on Nonlinear Theory and Applications*, pages 61–66.

Cichocki, A. and Unbehauen, R. (1994). *Neural Networks for optimization and Signal Processing.* John Wiley.

Cichocki, A. and Unbehauen, R. (1996). Robust neural networks with on-line learning for blind identification and blind separation of sources. *IEEE Trans. on Circuits and Systems - I: Fundamental Theory and Applications*, 43(21):894–906.

Cichocki, A., Unbehauen, R., and Rummert, E. (1994). Robust learning algorithm for blind separation of signals. *Electronics Letters*, 30(17):1386–1387.

Coenen, O. J.-M. D., Lee, T.-W., and Sejnowski, T. J. (1998). Is a factorial code constructed in the cerebellum? *in preparation.*

Cohen, M. and Andreou, A. (1992). Current-mode subthreshold mos implementation of the herault-jutten autoadaptive network. *IEEE J. Solid-State Circuits*, 27(5):714–727.

Comon, P. (1994). Independent component analysis – a new concept? *Signal Processing*, 36(3):287–314.

Comon, P. (1996). Contrasts for multichannel blind deconvolution. *Signal Processing Letters*, 3(7):209–211.

Cover, T. and Thomas, J., editors (1991). *Elements of Information Theory*, volume 1. John Wiley and Sons, New York.

Deco, G. and Brauer, W. (1995). Nonlinear higher-order statistical decorrelation by volume-conserving neural architectures. *Neural Networks*, 8(4):525–535.

Deco, G. and Obradovic, D. (1996). *An Information-Theoretic Approach to Neural Computing*. Springer Verlag, ISBN 0-387-94666-7.

DeGroot, M. (1986). *Probability and statistics*. Addison-Wesley Pub. Co., Mass.

Deller, J., Proakis, J., and Hansen, J. (1993). *Discrete-Time Processing of Speech Signals*. Prentice Hall, New Jersey.

Douglas, S. and Cichocki, A. (1997). Locally-adaptive networks for blind decorrelation. *IEEE Trans. on Signal Processing*, accepted.

Douglas, S., Cichocki, A., and Amari, S. (1997). Multichannel blind separation and deconvolution of sources with arbitrary distribiutions. In *Proc. IEEE Workshop on NNSP*, pages 436–445.

Ehlers, F. and Schuster, H. (1997). Blind separation of convolutive mixtures and an application in automatic speech recognition in noisy environment. *IEEE Transactions on Signal processing*, 45(10):2608–2609.

Everitt, B. (1984). *An introduction to latent variable*. Chapman and Hall, London.

Feder, M., Weinstein, E., and Oppenheim, A. (1993). Multi-channel signal separation by decorrelation. In *IEEE Trans. Speech and Audio Processing*, volume 1:4, pages 405–413.

Ferrara, E. (1980). Fast implementation of lms adaptive filters. *IEEE Transactions on Acoustics, Speech and Signal Processing*, 28(4):474–478.

Field, F. (1994). What is the goal of sensory coding? *Neural Computation*, 6:559–601.

Fisher, R. (1936). *The use of multiple measurements in taxonomic problem*. Annual Eugenics, 7, Part II, 179-188.

Freund, Y. (1995). Boosting a weak learning algorithm by majority. *Information and Computation*, 121:2.

Friedman, J. (1987). *Exploratory Projection Pursuit*, volume 82:397. Journal of the American Statistical Association.

Gaeta, M. and Lacoume, J.-L. (1990). Source separation without prior knowledge: the maximum likelihood solution. *Proc. EUSIPO*, pages 621–624.

Ghahramani, Z. (1994). Solving inverse problems using an em approach to density estimation. *Proceedings of the 1993 Connectionist Models Summer School*, pages 316–323.

Ghahremani, D., Makeig, S., Jung, T., Bell, A. J., and Sejnowski, T. J. (1996). Independent Component Analysis of simulated EEG Using a Three-Shell Spherical Head Model. *Technical Report INC 9601.*

Girolami, M. (1997a). An alternative perspective on adaptive independent component analysis algorithms. Technical report, issn 1461-6122, Department of Computing and Information Systems, Paisley University, Scotland.

Girolami, M. (1997b). Self-organizing artificial neural networks for signal separation. Ph.d. thesis, Department of Computing and Information Systems, Paisley University, Scotland.

Girolami, M. (1998). An alternative perspective on adaptive independent component analysis algorithms. *Neural Computation*, to appear.

Girolami, M. and Fyfe, C. (1997a). An extended exploratory projection pursuit network with linear and nonlinear anti-hebbian connections applied to the cocktail party problem. *Neural Networks Journal*, in press.

Girolami, M. and Fyfe, C. (1997b). Extraction of independent signal sources using a deflationary exploratory projection pursuit network with lateral inhibition. *I.E.E Proceedings on Vision, Image and Signal Processing Journal*, 14(5):299–306.

Girolami, M. and Fyfe, C. (1997c). Generalised independent component analysis through unsupervised learning with emergent bussgang properties. In *Proc. ICNN*, pages 1788–1891, Houston, USA.

Girolami, M. and Fyfe, C. (1997d). Stochastic ica contrast maximisation using oja's nonlinear pca algorithm. *International Journal of Neural Systems*, in press.

Gray, M., Sejnowski, T. J., and Movellan, J. (1998). A comparison of visual representations for speechreading. *IEEE Pattern Analysis and Machine Intelligence*, submitted.

Hateren, J. and Ruderman, D. (1998). Independent component analysis of video yields spatiotemporal filtering like in simple cells. *Technical Report.*

Haykin, S. (1991). *Adaptive filter theory.* Prentice-Hall.

Haykin, S. (1994a). *Blind Deconvolution.* Prentice Hall, New Jersey.

Haykin, S. (1994b). *Neural networks: a comprehensive foundation.* MacMillan, New York.

Hebb, D. O. (1949). *The Organization of Behavior: A Neurophysiological Theory.* Wiley, New York.

Herault, J. and Jutten, C. (1986). Space or time adaptive signal processing by neural network models. In in Denker J.S. (ed), editor, *Neural networks for computing: AIP conference proceedings 151*, New York. American Institute for physics.

Hermann, M. and Yang, H. (1996). Perspectives and limitations of self-organizing maps. In *ICONIP'96.*

Hertz, J., Krogh, A., and Palmer, R. (1991). *Introduction to the theory of neural computation.* Addison-Wesley Publishing Company.

Hillyard, S. and Picton, T. (1980). *Handbook of Phychology - The Nervous System V*, chapter Electrophysiology of cognition, pages 519–584. Book.

Hinton, G. and Ghahramani, Z. (1997). Generative models for discovering sparse distributed representations. Philosophical Transactions Royal Society B, 352:1177–1190.

Hochreiter, S. and Schmidhuber, J. (1998). Feature extraction through lococode. *Neural Computation*, to appear.

Hopfield, J. (1991). Olfactory computation and object perception. *Proc. Natl. Acad. Sci. USA*, 88:6462–6466.

Hyvaerinen, A. (1997). A family of fixed-point algorithms for independent component analysis. In *ICASSP*, pages 3917–3920, Munich.

Hyvaerinen, A. and Oja, E. (1997a). A fast fixed-point algorithm for independent component analysis. *Neural Computation*, 9:1483–1492.

Hyvaerinen, A. and Oja, E. (1997b). One-unit learning rules for independent component analysis. In *Advances in Neural Information Processing Systems 10*, pages 480–486.

Icart, S. and Gautier, R. (1996). Blind separation of convolutive mixtures using second and fourth order moments. In *Proc. ICASSP*, pages 3018–3021, Atlanta, GA.

Isbell, C. L. and Viola, P. (1998). Restructuring sparse digh dimensional data for effective retrieval. In *Advances in Neural Information Processing Systems 11*, volume submitted.

Jolliffe, I., editor (1986). *Principal Component Analysis*, volume 1. Springer-Verlag, New York.

Jones, M. and Sibson, R. (1987). What is projection pursuit. *The Royal Statistical Society*, A150:1–36.

Jung, T.-P., Humphries, C., Lee, T.-W., Makeig, S., McKeown, M., Iragui, V., and Sejnowski, T. J. (1998a). Extended ica removes artifacts from electroencephalographic recordings. *Advances in Neural Information Processing Systems 10*, in press.

Jung, T.-P., Humphries, C., Lee, T.-W., Makeig, S., McKeown, M., Iragui, V., and Sejnowski, T. J. (1998b). Removing electroencephalographic artifacts by blind source separation. *Journal of Psychophysiology*, submitted.

Jung, T.-P., Humphries, C., Lee, T.-W., McKeown, M., Iragui, V., Makeig, S., and Sejnowski, T. J. (1998c). Removing electroencephalographic artifacts: Comparison between ica and pca. *Proc. IEEE NNSP*, in press.

Jung, T.-P., Makeig, S., Bell, A. J., and Sejnowski, T. J. (1998d). Independent component analysis of electroencephalographic and event-related potential data. *In: P. Poon, J. Brugge, ed., Central Auditory Processing and Neural Modeling,*, pages 189–197.

Jung, T.-P., Makeig, S., Stensmo, M., and Sejnowski, T. J. (1996). Estimating alertness from the EEG power spectrum. *IEEE Transactions on Biomedical Engineering*, 44:60–69.

Jutten, C. and Herault, J. (1991). Blind separation of sources, part I: An adaptive algorithm based on neuromimetic architecture. *Signal Processing*, 24:1–10.

Kaliath, T. (1980). *Linear Systems*. Prentice Hall, New Jersey.

Karhunen, J. (1996). Neural approaches to independent component analysis and source separation. In *Proc. 4th European Symposium on Artificial Neural Networks*, pages 249–266, Bruges, Belgium.

Karhunen, J., Cichocki, A., Kasprzak, W., and Pajunen, P. (1997a). On neural blind separation with noise suppression and redundancy reduction. *Int. Journal of Neural Systems*, 8(2):219–237.

Karhunen, J., Hyvaerinen, A., Vigario, R., Hurri, J., and Oja, E. (1997b). Applications of neural blind separation to signal and image processing. In *ICASSP*, pages 131–134, Munich.

Karhunen, J. and Joutsensalo, J. (1994). Representaion and Separation of Signals using Nonlinear PCA Type Learning. *Neural Networks*, 7:113–127.

Karhunen, J., Oja, E., Wang, L., Vigario, R., and Joutsensalo, J. (1997c). A class of neural networks for independent component analysis. *IEEE Trans. on Neural Networks*, 8:487–504.

Karhunen, J., Wang, L., and Vigario, R. (1995). Nonlinear PCA type approaches for source separation and independent component analysis. In *Proc. ICNN*, pages 995–1000, Perth, Australia.

Koehler, B., Lee, T.-W., and Orglmeister, R. (1997). Improving the performance of infomax using statistical signal processing techniques. In *Proceedings International Conference on Artificial Neural Networks*, volume 535-540.

Kohonen, T. (1989). *Self-Organization and Associative Memory*. Springer-Verlag, Berlin.

Kosko, B. (1992). *Neural Networks for Signal Processing*. Prentice Hall, New Jersey.

Kwong, K., Belliveau, J., Chesler, D., Goldberg, I., Weisskoff, R., and Poncelet, B. (1992). Dynamic magnetic resonance imaging of human brain activity during primary sensory stimulation. *Proc Natl Acad Sci U S A*, 89(12):5675–9.

Lambert, R. (1996). Multichannel blind deconvolution: Fir matrix algebra and separation of multipath mixtures. Thesis, University of Southern California, Department of Electrical Engineering.

Lambert, R. and Bell, A. J. (1997). Blind separation of multiple speakers in a multipath environment. In *ICASSP*, pages 423–426, Munich.

Lambert, R. and Nikias, C. (1995a). Polynomial matrix whitening and application to the multichannel blind deconvolution problem. In *IEEE Conference on Military Communications*, pages 21–24, San Diego, CA.

Lambert, R. and Nikias, C. (1995b). A sliding cost function algorithm for blind deconvolution. In *29th Asilomar conf. on Signals Systems and Computers*, pages 177–181.

Laubach, M. and Nicolelis, M. (1998). The independent components of neural populations are cell assemblies. *Advances in Neural Information Processing Systems 11*, submitted.

Lee, T.-W., Bell, A. J., and Lambert, R. (1997a). Blind separation of convolved and delayed sources. In *Advances in Neural Information Processing Systems 9*, pages 758–764. MIT Press.

Lee, T.-W., Bell, A. J., and Orglmeister, R. (1997b). Blind source separation of real-world signals. In *Proc. ICNN*, pages 2129–2135, Houston, USA.

Lee, T.-W., Girolami, M., Bell, A. J., and Sejnowski, T. J. (1998a). A unifying framework for independent component analysis. *International Journal on Mathematical and Computer Models*, in press.

Lee, T.-W., Girolami, M., and Sejnowski, T. J. (1998b). Independent component analysis using an extended infomax algorithm for mixed sub-gaussian and super-gaussian sources. *Neural Computation*, in press.

Lee, T.-W., Koehler, B., and Orglmeister, R. (1997c). Blind separation of nonlinear mixing models. In *IEEE NNSP*, pages 406–415, Florida, USA.

Lee, T.-W., Lewicki, M. S., Girolami, M., and Sejnowski, T. J. (1998c). Blind source separation of more sources than mixtures using overcomplete representations. *IEEE Signal Processing Letters*, submitted.

Lee, T.-W., Lewicki, M. S., and Sejnowski, T. J. (submitted 1998d). Unsupervised classification with non-gaussian mixture models using ica. In *Advances in Neural Information Processing Systems 11*. MIT Press.

Lee, T.-W. and Sejnowski, T. J. (1997). Independent component analysis for sub-gaussian and super-gaussian mixtures. In *4th Joint Symposium on Neural Computation*, volume 7, pages 132–139. Institute for Neural Computation.

Lee, T.-W., Ziehe, A., Orglmeister, R., and Sejnowski, T. J. (1998e). Combining time-delayed decorrelation and ica: Towards solving the cocktail party problem. In *Proc. ICASSP*, volume 2, pages 1249–1252, Seattle.

Lewicki, M. and Olshausen, B. (1998). Inferring sparse, overcomplete image codes using an efficient coding framework. *J. Opt.Soc.A,. A: Optics, Image Science and Vision*, submitted.

Lewicki, M. and Sejnowski, T. J. (1998a). Coding time-varying signals using sparse shift-invariant representations. In *Advances in Neural Information Processing Systems 11*, volume submitted.

Lewicki, M. and Sejnowski, T. J. (1998b). Learning nonlinear overcomplete represenations for efficient coding. In *Advances in Neural Information Processing Systems 10*, volume in press.

Lewicki, M. and Sejnowski, T. J. (1998c). Learning overcomplete representations. *Neural Computation*, submitted.

Li, S. and Sejnowski, T. J. (1995). Adaptive separation of mixed broad-band sound sources with delays by a beamforming Hérault-Jutten network. *IEEE Journal of Oceanic Engineering*, 20(1):73–79.

Lin, J. and Cowan, J. (1997). Faithful Representation of separable input distributions. *Neural Computation*, 9:6:1305–1320.

Lin, J., Grier, D., and Cowan, J. (1997). Feature extraction approach to blind source separation. *IEEE Workshop on Neural Networks for Signal Processing*, pages 398–405.

Linsker, R. (1989). An application of the principle of maximum information preservation to linear systems. In *Advances in Neural Information Processing Systems 1*.

Linsker, R. (1992). Local synaptic learning rules suffice to maximize mutual information in a linear network. *Neural Computation*, 4:691–702.

Lizhong, W. and Moody, J. (1997). Multi-effect decompositions for financial data modeling. In *Advances in Neural Information Processing Systems 9*, pages 995–1001. MIT Press.

MacKay, D. (1995). Probable networks and plausible predictions - a review of practical bayesian methods for supervised neural networks. Book chapter, University of Cambridge, Cavendish Lab.

MacKay, D. (1996). Maximum likelihood and covariant algorithms for independent component analysis. Report, University of Cambridge, Cavendish Lab.

MacKay, D. (1998). *Information Theory, Inference and Learning Algorithm*. Internet, `http://wol.ra.phy.cam.ac.uk/mackay/itprnn/#book` .

Makeig, S., Bell, A. J., Jung, T., and Sejnowski, T. J. (1996). Independent Component Analysis of Electroencephalographic Data. *Advances in Neural Information Processing Systems 8*, pages 145–151.

Makeig, S. and Inlow, M. (1993). Changes in the eeg spectrum predict fluctuations in error rate in an auditory vigilance task. In *Society for Psychophysiology*, volume 28:S39.

Makeig, S., Jung, T., Bell, A. J., Ghahremani, D., and Sejnowski, T. J. (1997). Blind Separation of Event-related Brain Response into spatial Independent Components. *Proc. of the National Academy of Sciences*, 94:10979–10984.

Manoach, D., Schlaug, G., Siewert, B., Darby, D., Bly, B., Benfield, A., and Edelman, R. (1997). Prefrontal cortex fmri signal changes are correlated with working memory load. *Neuroreport*, 8(2):545–549.

Matsuoka, K., Ohya, M., and Kawamoto, M. (1995). A neural net for blind separation of nonstationary signals. *Neural Networks*, 8(3):411–419.

McKeown, M., Makeig, S., Brown, G., Jung, T.-P., Kindermann, S., Bell, A. J., and Sejnowski, T. J. (1998a). Analysis of fmri by blind separation into independent spatial components. *Human Brain Mapping*, 6:1–31.

McKeown, M., Makeig, S., Brown, G., Jung, T.-P., Kindermann, S., Lee, T.-W., and Sejnowski, T. J. (1998b). Spatially independent activity patterns in functional magnetic resonance imaging data during the stroop color-naming task. *Proceedings of the National Academy of Sciences*, 95:803–810.

Mendel, J. (1990). *Maximum Likelihood Deconvolution.* Springer-Verlag, Berlin.

Merz, C. and Murphy, P. (1998). UCI repository of machine learning databases.

Molgedey, L. and Schuster, H. (1994). Separation of independent signals using time-delayed correlations. *Phys. Rev. Letts*, 72(23):3634–3637.

Moulines, E., Cardoso, J.-F., and Cassiat, E. (1997). Maximum likelihood for blind separation and deconvolution of noisy signals using mixture models. In *Proc. ICASSP'97*, volume 5, pages 3617–3620, Munich.

Murata, N., Ikeda, S., and Ziehe, A. (1998). An approach to blind source separation based on temporal structure of speech signals. *IEEE Transactions on signal processing*, submitted.

Murata, N., Mueller, K.-R., Ziehe, A., and Amari, S. (1997). Adaptive on-line learning in changing environments. In *Advances in Neural Information Processing Systems 9*, pages 599–605. MIT Press.

Nadal, J.-P. and Parga, N. (1994). Non linear neurons in the low noise limit : a factorial code maximizes information transfer. *Network*, 5:565–581.

Nadal, J.-P. and Parga, N. (1997). Redundancy reduction and independent component analysis: Conditions on cumulants and adaptive approaches. *Neural Computation*, 9:1421–1456.

Nadal, J.-P. and Parga, N. (1998). Blind source separation with time dependent mixtures. *submitted to Signal Processing.*

Nelson, A. and Wan, E. (1997a). Dual kalman filtering methods for nonlinear prediction, smoothing, and estimation. In *Advances in Neural Information Processing Systems 9*, pages 793–800. MIT Press.

Nelson, A. and Wan, E. (1997b). Neural Speech Enhancement Using Dual Extended Kalman Filtering. In *Intern. Conference on Neural Networks*, pages 2171–2176.

Nguyen-Thi, H.-L. and Jutten, C. (1995). Blind source separation for convolutive mixtures. *Signal Processing*, 45(2).

Nobre, A., Sebestyen, G., Gitelman, D., Mesulam, M., Frackowiak, R., and Frith, C. (1997). Functional localization of the system for visuospatial attention using positron emission tomography. *Brain*, 120:513–533.

Nunez, P. (1981). *Electric Fields of the Brain.* Oxford, New York.

Oja, E. (1982). A simplified neuron model as a principal component analyzer. *Journal of Mathematical Biology*, 53:267–273.

Oja, E. (1997). The nonlinear pca learning rule in independent component analysis. *Neurocomputing*, 17:25–45.

Oja, E. and Karhunen, J. (1995). Signal separation by nonlinear hebbian learning. In *ICNN*, pages 83–87, Perth, Australia. IEEE.

Olshausen, B. and Field, D. (1996). Emergence of simple-cell receptive field properties by learning a sparse code for natural images. *Nature*, 381:607–609.

Oppenheim, A. and Schafer, R. (1989). *Discrete-Time Signal Processing*. Prentice Hall, New Jersey.

Pajunen, P. (1996). Nonlinear independent component analysis by self-organizing maps. Technical report, Proc. ICANN.

Pajunen, P. (1997). Blind separation of binary sources with less sensors than sources. In *Proc. ICNN*, pages 1994–1997, Houston, USA.

Pajunen, P. and Karhunen, J. (1997). A maximum likelihood approach to nonlinear blind source separation. In *ICANN*, pages 541–546, Lausanne.

Papoulis, A., editor (1990). *Probability and Statistics*, volume 1. Prentice Hall, New Jersey.

Pearlmutter, B. and Parra, L. (1996). A context-sensitive generalization of ICA. In *International Conference on Neural Information Processing*, pages 151–157.

Pearlmutter, B. and Parra, L. (1997). Maximum likelihood blind source separation: A context-sensitive generalization of ICA. In *Advances in Neural Information Processing Systems 9*, pages 613–619. MIT Press.

Pearson, K. (1894). Contributions to the mathematical study of evolution. *Phil. Trans. Roy. Soc. A*, 185(71).

Pham, D.-T. and Garrat, P. (1997). Blind separation of mixture of independent sources through a quasi-maximum likelihood approach. *IEEE Trans. on Signal Proc.*, 45(7):1712–1725.

Pham, D.-T., Garrat, P., and Jutten, C. (1992). Separation of a mixture of independent sources through a maximum likelihood approach. In *Proc. EUSIPCO*, pages 771–774.

Platt, J. and Faggin, F. (1992). Networks for the separation of sources that are superimposed and delayed. In *Advances in Neural Information Processing Systems 4*, pages 730–737.

Rabiner, L. and Juang, B.-H., editors (1993). *Fundamentals of Speech Recognition*, volume 1. Prentice Hall, New Jersey.

Rojas, R. (1996). *Neural Networks - A Systematic Introduction*. Springer-Verlag, Berlin.

Roth, Z. and Baram, Y. (1996). Multidimensional density shaping by sigmoids. *IEEE Trans. on Neural Networks*, 7(5):1291–1298.

Sanger, T. (1989). Optimal unsupervised learning in a single layer linear feedforward neural network. *Neural Networks*, 12:459–473.

Shannon, C. E. (1948). A mathematical theory of communication. *Bell Sys. Tech. Journal*, 27:379–423, 623–659.

Simoncelli, E., Freeman, W., Adelson, E., and J., H. (1992). Shiftable multiscale transforms. *IEEE Transactions on Info. Theory*, 38:587–607.

Smaragdis, P. (1997). Information theoretic approaches to source separation. Master thesis, Massachusetts Institute of Technology, MAS Department.

Sorouchyari, E. (1991). Blind separation of sources, part III: Stability analysis. *Signal Processing*, 24(1):21–29.

Stuart, A. and Ord, J. (1987). *Kendall's Advanced Theory of Statistic, 1, Distribution Theory.* John Wiley, New York.

Taleb, A. and Jutten, C. (1997). Nonlinear source separation: The post-nonlinear mixtures. In *ESANN*, pages 279–284.

Toga, A. and Mazziotta, J. (1996). *Brain Mapping, The Methods.* Academic Press, San Diego.

Tong, L., Soon, V., Huang, Y., and Lui, R. (1991). A necessary and sufficient condition for the blind separation of memoryless systems. *Proc. IEEE ISCAS*, pages 1–4.

Torkkola, K. (1996a). Blind separation of convolved sources based on information maximization. In *IEEE Workshop on Neural Networks for Signal Processing*, pages 423–432, Kyoto, Japan.

Torkkola, K. (1996b). Blind Source Separation of Delayed Sources Based on Information Maximization. In *ICASSP*, pages 3509–3512.

Torkkola, K. (1998). Blind separation of radio signals in fading channels. In *Advances in Neural Information Processing Systems 10*, volume in press. MIT Press.

Turk, M. and Pentland, A. (1991). Eigenfaces for recognition. *Journal of Cognitive Neuroscience*, 3(1):71–86.

Vapnik, V., Golowich, S., and Smola, A. (1997). Support vector method for function approximation, regression and estimation, and signal processing. In *Advances in Neural Information Processing Systems 9*, pages 281–287. MIT Press.

Vigario, R., Hyvaerinen, A., and Oja, E. (1996). Ica fixed-point algorithm in extraction of artifacts from eeg. In *IEEE Nordic Signal Processing Symposium*, pages 383–386, Espoo, Finland.

Vittoz, A. and Arreguit, X. (1989). Cmos integration of herault-jutten cells for separation of sources. *Analog VLSI implementation of neural systems*, pages 57–84.

Widrow, B. and Stearns, S. (1985). *Adaptive Signal Processing.* Prentice Hall, New Jersey.

Xu, L. (1993). Least mse reconstruction: A principle for self organizing nets. *Neural Networks*, 6:627–648.

Xu, L., Cheung, C., Yang, H., and Amari, S. (1997). Maximum equalization by entropy maximization and mixture of cumulative distribution functions. In *Proc. of ICNN'97*, pages 1821–1826, Houston.

Yang, H. and Amari, S. (1997). Adaptive on-line learning algorithms for blind separation - maximum entropy and minimum mutual information. *Neural Computation*, 9:1457–1482.

Yang, H., Amari, S., and Cichocki, A. (1997). Information back-propagation for blind separation of sources from non-linear mixtures. In *Proc. of ICNN*, pages 2141–2146, Houston.

Yellin, D. and Weinstein, E. (1994). Criteria for multichannel signal separation. *IEEE Transactions on Signal Processing*, 42(8):2158–2168.

Yellin, D. and Weinstein, E. (1996). Multichannel signal separation: Methods and analysis. *IEEE Transactions on Signal Processing*, 44(1):106–118.

About the author

Te-Won Lee was born in June 1969 in Chungnam, Korea. He received his diploma degree in March 1995 and his Ph.D. degree in October 1997 in electrical engineering with highest honor from the University of Technology Berlin. He was a visiting graduate student at the Institute Nationale de Polytechnique de Grenoble, the University of California at Berkeley and the Carnegie Mellon University. From 1995 to 1997 he was a Max-Planck Institute fellow. Since 1997, he is a research associate at the Computational Neurobiology Laboratory at The Salk Institute where he is interested in biomedical signal processing applications and unsupervised learning algorithms. His current focus are ICA applications and unsupervised classification algorithms using ICA.

Index